Developing a Social Psychology of Monkeys and Apes

John Chadwick-Jones
Institute of Psychology
University of Fribourg
Switzerland

Psychology Press
a member of the Taylor & Francis group

To Araceli, Diane, John, and Andrew

Psychology Press, Publishers
27 Church Road
Hove
East Sussex, BN3 2FA
UK

British Library Cataloguing in Publication Data
A catalogue record for this title is available from the British Library

ISBN: 0-86377-820-8 (hbk)

Typeset by Graphicraft Limited, Hong Kong
Printed and bound in the UK by Biddles Ltd, Guildford and King's Lynn

Contents

Acknowledgements

I would like to acknowledge especially the unfailing collegiality and support of Hans-Dieter Schneider, Oswald Huber, and Meinrad Perrez.

I thank Robert Hinde for his encouragement of my interest in primate social activities and for his criticisms of my working papers that were subsequently published in the late 1980s. I value his viewpoint that the study of primates (human and nonhuman) must involve both short-term causal explanations and long-term evolutionary ones.

My heartfelt thanks are expressed to Vernon Reynolds for countless discussions of topics connected with the social actions of monkeys and apes, and for his frequent advice. Robin Dunbar answered my questions on a number of occasions and gave his time to discuss some of the theoretical issues in this area. I am grateful to Michael Simpson for his collaboration and suggestions for observational studies of rhesus monkeys that I carried out at the University of Cambridge. I was privileged also to discuss shared interests with primatologists at the primate research centres in Cambridge, Zurich, and Madrid.

My thanks are expressed to Gwyneth Doherty-Sneddon for commentaries on some sections of the draft manuscript and to Vernon Reynolds for commentaries on the completed manuscript. Two anonymous reviewers read the manuscript and offered criticisms for which I am deeply indebted to them. All their perceptive comments led to improvements in the manuscript but responsibility for the views and conclusions in the text is, of course, entirely mine.

Barbara Smuts was extremely generous in making available the extracts from her unpublished field reports, samples of which are presented in Figs. 5.2 and 8.1. Thanks are due to Les Barden of the University of Cambridge for his careful work on the line drawings of the figures in Chapters 5 and 8.

Figure 1.1, "A sociogram of brown howler monkeys" from A.G. Chiarello (1995) *Grooming in six brown howler monkeys*, is reproduced by permission of *The American Journal of Primatology*. Tables 7.1 and 7.2 are reproduced from V. Reynolds (1976) *The origins of a behavioural vocabulary* by permission of *The Journal for the Theory of Social Behaviour*.

John Chadwick-Jones
University of Fribourg

CHAPTER ONE

Introduction: Social Psychology and Primates

USING A SOCIAL PSYCHOLOGICAL APPROACH

Researchers from a variety of disciplines have studied the social behaviour of monkeys and apes, and usually they have done so because of their interest in the evolutionary origins of primates (including humans). Many have undertaken observational studies in the field, or have devised experiments to test some aspect of the mental process of relating to others. Often they have taken an approach that is similar to that of social psychologists who study interpersonal communication. In this case, there has been some influence on primatology from the discipline of social psychology but, at the same time, the work of primatologists has begun increasingly to influence views and interpretations of what humans do or plan to do. For example, in attempts to explain the apparently unique human emotion of embarrassment, studies have drawn on parallels with the appeasement signals of monkeys and apes: the smiles, gaze aversion, lowering of the head and body, self-touching, all of which could suggest a connection with the management of embarrassing situations by humans (Keltner & Buswell, 1997). Writers who argue the case for an "evolutionary social psychology" have drawn attention to the importance of both the proximate (immediate causes, context) and ultimate (evolutionary, functional) sources of explanation if we are to reach a fuller understanding of social relationships. Kenrick and Simpson (1997), advancing the argument for an evolutionarily valid social psychology, insist that social events, behaviour, and relationships, will not be understood fully unless we refer to their likely evolutionary significance and origins (see also Buss, 1995). One possible way ahead might be to use the framework of evolutionary theory as an organizing influence for the fragmented mini-theories of social psychology, as Kenrick and Simpson suggest (1997, p.15), thus offering an opportunity "to go beyond a proximal level of analysis if we want to understand human social behaviour from a thoroughly scientific perspective." While recognizing this as an eventual objective, the aim of this book is to give prime attention to proximate causes and to extend our knowledge of them.

In studies of nonhuman primates, an emphasis on social behaviour and relationships is found in the work of a very productive group of researchers, although traditionally primate studies also deal with anatomical, physiological, reproductive and ecological topics. Much of the contemporary research effort has been directed to ecology (a field that is complementary to social psychology) and to the neurosciences (Fady, 1990). However, in zoology, there has been a great deal of interest in social stratagems that can be shown to improve the chances of survival or reproduction. This interest emerges in references to "receiver psychology" or "how animals receive signals" as an important area for research: according to Krebs (1991, p. 69), "we have to know about the psychology of the animals who are responding to the signal" (see also Guilford & Dawkins, 1991; Krebs & Davies, 1993). Recently, there have been strong influences on studies of monkeys and apes from studies of human cognition (Whiten & Byrne, 1997).

The approach that is put into practice in the following pages draws attention to the social skills and tactics that monkeys and apes express in abundance. If, through this approach, we can achieve a better understanding of the Order: Primates, it will only be because of the work of the contemporary researchers whose studies are discussed in this book and who have provided the detailed descriptions and appraisals of what monkeys and apes do socially.

Already, researchers in primatology have used methods (like sociometry and the analysis of gestures or facial expressions) that derive from psychology and sociology. They often interpret what they observe by concepts that are familiar to social psychologists: is a gesture, for example, an expression of social approval? Does a session of grooming predict a future social exchange? We will discuss the answers to such questions at length in later chapters.

The social psychology of primates has become a shared domain of several disciplines, each with an interest in social interaction as a principal focus of study. There is, in this interest, no issue of one discipline encroaching on another. Studies of social interaction in monkeys and apes have already produced a vast amount of information collected by researchers working from university departments of zoology, anthropology, or psychology. I certainly would not suggest that a social psychological perspective, by its focus on face-to-face interactions, should supplant some other discipline, but rather that this perspective can make its contribution to an interdisciplinary enterprise. Social psychologists, for their part, can make use of the many studies by primatologists, ethologists, anthropologists, and psychologists that I discuss in the following chapters, as a means to inform their own perspective on human studies.

CONVERSATIONS: A SOCIAL PSYCHOLOGICAL FOCUS

Communication among monkeys and apes is an ideal subject for social psychological treatment because the signals exchanged occur in sequences analogous to human conversations. Recent studies have introduced significant discussions of the meaning of signals, and of the proximate intentions of callers. It is important to consider this rapidly moving and exciting area of new research.

The main themes in the following pages centre on the detailed analysis of the social contacts of nonhuman primates. The concepts and methods used in the discipline of social psychology are many, but textbooks usually focus on an individual's actions towards others in close relationships and refer to communication at this level of face-to-face contact. If we want to use a similar approach to the study of nonhuman primates, then the very first problem to acknowledge is that human communication by language does not have direct parallels here. So much of social psychology (theory and methodology) involves the use and the meanings of language. We cannot transfer this to nonhuman primates. Nevertheless much of human communication is also nonverbal, by mutual gaze, posture, gesture, touch, and is related to the signalling of monkeys and apes. As well as this, some of the vocal signals of monkeys have characteristics that justify considering them as a quasi-linguistic stage of communication, because of the semantic and pragmatic uses of these signals.

My aim is to highlight the work of the researchers who have provided detailed descriptions. In Chapters 2 and 3, I discuss some definitions and general theoretical questions to be confronted in studies of primate communication. I will examine the progress of research on facial expressions in Chapter 4, on gestures and touching in Chapter 5, and on vocal signals in Chapters 6 and 7. Many field workers have studied these topics with the functional (evolutionary) significance of the behaviour in mind, but they have, in fact, focused on the detail of social interactions and their possible meanings *in situ*. An example will be found in Chapter 8 in the discussion of infant-buffering, the enigmatic stratagem of male baboons—challenging each other while carrying an infant.

It is not my aim only to review the literature on these topics but to select examples that show how signals are used in a social context. I will refer to earlier work as well as to recent studies, so that the reader can get a fair idea of historical developments in the study of a particular type of signal. An emphasis on earlier and classic studies is important because these studies generated ideas or contributed refinements to the explanations used. The interpretations of researchers are very often the critical issue in the chapters that follow.

After the chapters on aspects of communication in their social context, I will go on to discuss the more general topics of social exchange, sexual behaviour, and dominance hierarchies (Chapters 9, 10, 11), with particular reference to the development of new explanations in current studies. Appendix 1 gives details of primate species, their social structures and characteristic forms of group living. Appendix 2 discusses biological affinities among nonhuman primates and humans.

I have taken a certain perspective, I will show how the social psychological focus on the actions of monkeys and apes is of value because of what it reveals about primates—in particular, that nonhuman primates are on a continuum with humans, whether it be in the social uses of gesture, of touch, of vocal sounds, or of facial expressions, or how these are used jointly to achieve a social objective. Very often this subject matter seems elusive, there is not sufficient detail or one would like to see a sequence of actions described for a longer period. Some of the recent studies, rich in detail, assure me that much more will be known and understood, very soon, by those who follow in this area.

I have been a field worker observing human primates but I write here about researchers' work as observers of monkeys and apes. The procedures and ethos of field and experimental studies are in general the same, whatever the species of primate that is being studied. To that extent, my work in bringing together the large variety of studies that comprise a developing social psychology has been made easier. For the first time, these many studies on a range of topics and from diverse sources will be discussed together under one rubric.

In the following pages the social interactions of monkeys and apes will be the focus of study. Questions will be raised about how, when, or why certain social contacts take place and I will try to identify immediate causes and effects in a given situation. The resulting explanations will be at the proximate level, distinct from but complementary to evolutionary theory. The reason for taking such an approach is to find out more about what monkeys and apes are capable of doing in social situations. Do they act as if they have purposes or even plans in pursuing a certain outcome? Do their actions give evidence of thinking, possibly with a particular objective in mind? These questions are discussed in the following chapters.

Researchers have given us a rich account of the communication and social stratagems of monkeys and apes. Detailed reporting of micro-signals similar to human nonverbal communication, allows us to start to build up a picture of the social psychology of primates. The work of contemporary researchers, in this area of rapid advances, is extending our knowledge, year by year, of the social behaviour of nonhuman primates. If writers in the discipline of social psychology were to recognize the importance of nonhuman primate studies, there would be at least one chapter on this topic in every introductory textbook in the discipline. But we are still a long way from that at the moment.

SUITABLE TOPICS: RECONCILIATION

Some topics seem obviously suitable for a social psychological approach, such as the process of reconciliation expressed in friendly signals exchanged between individuals shortly after fighting or threatening each other. These episodes are of interest because of the clear intentional and distinctive use of gestures such as the outstretched arm and open hand used by chimpanzees (de Waal, 1989a). Following on from de Waal's pioneering work, the reconciliation concept has been taken into experimental situations and the reciprocal use and meaning of signals have had much attention.

The current research achievement in identifying the process of reconciliation is a significant one (de Waal, 1996a, b) with gradual accumulation of observations. For instance, after reconciliations among Barbary macaques (*Macaca sylvanus*), a decrease in nervous self-scratching was reported (Aureli, 1994): this was a useful indicator that social tensions were reduced. In an experiment among long-tailed or Java macaques (*Macaca fascicularis*) reconciliations were made a condition for coordinated feeding. Thus, harmonious relationships became more valuable, and the more valuable the relationship, the more reconciliation efforts were made (Cords & Thurnheer, 1993). Research has also confirmed that low-amplitude vocal sounds are an important part of the reconciliation process (Cheney, Seyfarth, & Silk, 1995a). In chimpanzees, reconciliation has been shown to be important for the maintenance of stable relationships after competition (de Waal & Aureli, 1996). Each one of these studies has been a step on the way to giving fuller attention to proximate causes.

THE FOUR WHYS OF TINBERGEN, AND SOCIAL BEHAVIOUR

Ethologists have been very influential in providing theoretical guidelines for the study of social behaviour (Hinde, 1982). Nikko Tinbergen (1963), in his famous appraisal of the aims of ethology, mentioned four major questions (the "four whys"): those touching on (1) immediate causation (proximate factors), (2) survival (function), (3) learning (development), and (4) evolution (reproductive success). Tinbergen argued for equal consideration of them all. In the following chapters, observational studies of proximate behaviour have priority for discussion. A social psychological approach aims to consider the how, when, how often, and why of the joint actions of individuals. Social actions as they occur in the natural habitat will often be the focus of our interest.

The social psychology of nonhuman primates is treated here as a relatively autonomous area in data-collection and theory, but I do not want to overemphasize the degree of autonomy. There is a great deal of shared ground with other levels of explanation; for example, with behavioural ecology. Social relationships—even if studied at the level of currently happening social events (the proximate level)—may also be understood at another level of theory as functional and related to biological survival. Social events are not completely in a separate explanatory compartment called causal (i.e. immediate or proximate); they should be also understood as interrelated with biological structures.

SOCIAL PSYCHOLOGICAL METHODS

Social psychology, as an academic discipline, deals with how and why individuals behave, think, and feel in situations involving other persons (Baron & Byrne, 1991). The similarities between monkeys, apes, and humans, as primates, suggest that this frame of reference might be as suitable for monkeys and apes as for humans. An apparent obstacle in applying this definition of social psychology to monkeys and apes is that the *think* and *feel* parts of the definition are not *accessible*, as in the case of humans, *through language*, but this still leaves an immense range of visual and vocal signals as a basis for inferences about thinking and feeling. Observers may agree about an individual's actions, but what a monkey or ape thinks in social situations can only be known by inference. The process of drawing conclusions about thinking or feeling in nonhuman primates is therefore both a challenge and a source of ambiguity.

Field workers in the tradition of ethology have compiled lists of signals (sounds, gestures) of various kinds, for different species of monkeys and apes. The compilation of lists of signals is a necessary beginning. Having started from taxonomies or vocabularies of this kind, it is now necessary to focus on how these signals combine to produce messages from sender to recipient and back again. Social interactions are the entry point for an empirical social psychology of monkeys and apes.

SOCIOGRAMS

Sociometry is a technique that consisted originally, in social psychology, of asking group members to list individuals in their group whom they would choose as friends. Adding the choices produces a popularity index of the most frequently chosen, and the isolates (chosen by and choosing no-one) or members liked least by others.

Sociometry first appeared in primate studies with the use of sociograms in studies of gibbons (*Hylobates* species), to represent spatial and time associations (Carpenter, 1945). Later on, the frequencies of grooming between rhesus monkeys (*Macaca mulatta*) were used as the equivalent of a sociometric choice, a kind of voting with the hands. Sociograms showed which were the most sought-after grooming partners (Sade, 1972). The sociograms, based on observations made over periods of a year or so, showed that monkeys associated most with their closest kin; demonstrating the strength of kinship bonds.

The technique has been resilient in uncovering relationships within groups and in showing the behavioural impact of dominance relationships and of alliances.

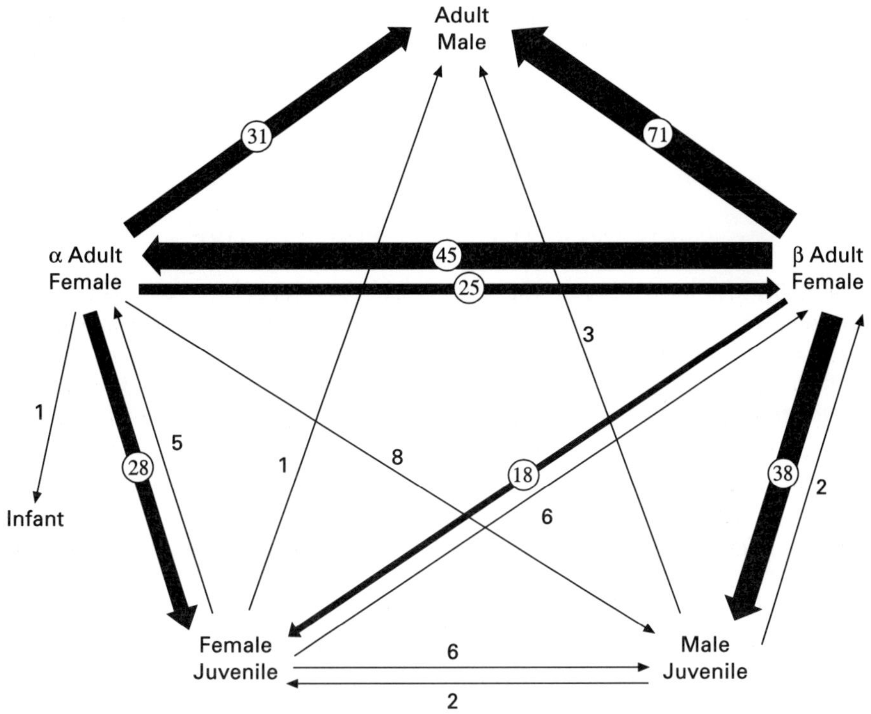

FIG. 1.1. Sociogram of brown howler monkeys (*Alouatta fusca*) (From Chiarello, 1995, reprinted with permission).

The example in Fig. 1.1 shows a sociogram of grooming between five brown howler monkeys. It is based on 290 grooming bouts observed over a year. The thickness of arrows is proportional to the amount of grooming. The figures inside the circles are the corresponding totals of grooming bouts initiated or received by each monkey. Here, the adult male received most of the grooming, although he never groomed others.

A very complex but inventive use of the sociogram occurred in a study of playmate relationships in a troop of Japanese monkeys (*Macaca fuscata*) (Koyama, 1985). A total of 5902 play bouts were analyzed among 1–5-year-old females and males to produce the sociogram. The influences of kinship, sex, and age on the frequency of play could be explored by this method. Kin relationships were influential for both sexes but, as the sociogram showed, males tended to play more with unrelated individuals of the same age. Females played more with related partners of different ages. These patterns gave the basis for a number of conjectures about their adaptive significance: for instance, the likelihood that bonds formed through play in early years led to alliances at later ages.

Besides using sociograms, we should be looking to social psychology for the analysis of sequences of actions. This type of analysis will give us a basis for making inferences about the purposes guiding actions, the tactics used to achieve them, and about the nature of the thinking involved. Between social psychology and primatology there is a great deal of shared methodology, especially as found in observational primate studies. To develop a social psychology of monkeys and apes we will need to use observational methods that yield as much detail as possible, focusing always on reciprocal actions.

AN EVOLUTIONARY PERSPECTIVE

The subject of the following chapters will be the actions of monkeys and apes in communicating with others of their species. How do we fit a social psychological focus into an evolutionary framework? An evolutionary standpoint refers to the huge time spans required for the process of evolutionary change. It is hard to see how modes of communication figure in this process because there are no records of their past development. We can only try to compare present-day species and then make some informed guesses on likely evolutionary changes. Modes of communication are important because natural selection has operated through what is usually termed behaviour in particular environments. The adaptiveness and survival of species depends on such behaviour as reproducing and foraging, but this behaviour involves acting socially with other females and males. Social interaction must therefore be vitally important in evolution.

It is important to study the nature of social interaction in as much detail as has been provided up to now by research studies. The studies that are discussed in the following chapters emphasize the communication modes used by monkeys and apes in the wild. Degrees of flexibility of communication sequences will be a main topic of interest. The less stereotyped that these sequences are found to be, the more interest they carry for us, because they show how far there is adaptability among monkeys and apes in communicating to achieve social goals. One action leading to the next means that the effects of each action are monitored and influence the next action approaching the goal. In fact there may be a definite pause after an action has been made in anticipation of its effect (Tomasello et al., 1994). Actions are tried out, from a repertoire that can be varied. The degrees of variety and flexibility of actions are only partly known. We need to know more about how interactions happen and how signals are combined to achieve their social purpose.

Across the range of monkeys' and apes' social behaviour, the methodology used by researchers to study cognition and communication has been effective and inventive. We will proceed now to explore this body of research as far as it can take us.

CHAPTER TWO

Intentionality, Deception, and Social Intelligence

INTENTIONALITY

The social actions of monkeys and apes are our chosen subject matter, to be described in as much detail as possible. In studying apparent causes and effects, many researchers have understood these actions as purposive, reflecting intentions to influence others. Definitions of intentions are varied and are much debated. A major distinction is often made between intentions in humans, implying consciousness of purpose, and simple intentions in nonhuman primates. In the latter case, awareness of purpose is not assumed although monkeys and apes still have purposes in the sense of having objectives or goals. Actions may be interpreted as being mainly thought-controlled or behaviour-controlled, that is, controlled by reflexes. In studies of the social actions of nonhuman primates this can be an interesting but unresolved question. Often we just do not know the answer but the cumulative evidence now supports the view that, especially in apes, much social behaviour is thought-controlled. The preliminary survey of some current issues in this chapter will advance this viewpoint and it will be further argued throughout the chapters to follow.

We can take as our starting assumption for a social psychology of primates, that monkeys and apes act *as if* there was intention to achieve a given objective. This, as a working hypothesis, will allow us to get close to the actions to be observed, to report them in all their variety. There are many other animals, besides primates, that have complex abilities, including the larger-brained birds. But we should be more ready in the case of monkeys and apes to find evidence of thinking because of their similarities to humans in brain structure and often in behaviour (Povinelli & Preuss, 1995; see also Appendix 2).

The *as if* proposition has been called mock anthropomorphism, a legitimate approach if it helps us understand ultimate functions of behaviour but not if used to explain behaviour in terms of intentions or mental images of the goal (Kennedy, 1992). This objection pointed to the risk, in the *as if* proposition, of assuming that the behaviour had been explained in a proximately causal sense. According to Kennedy, proposing that animals have intentions is not an explanation but only an hypothesis, and it can be an error to attribute an action to purpose, as a cause, because this makes an outcome into an antecedent. He considered it important to avoid attributing intention to social actions by hindsight, making mentalistic attributions. However, these arguments fall far short of any constructive approach to the accumulating evidence on monkeys' and apes' social stratagems and reflect, as Bekoff (1995) has noted, an over-simple approach.

On the positive side, the *as if* assumption serves to encourage a closer examination of what monkeys and apes do socially. Some researchers have assumed no more than a desire psychology, recognizing that nonhuman primates have wants and desires but leaving aside the possibility that they have mental images of future goals. They would argue that nonhuman primates pursue certain objectives but do not necessarily imagine them. The reports on chimpanzee problem-solving suggest, however, that this species may well do so.

There have been several obstacles to the study of mental capacities in nonhuman primates. The influence of behaviourist ideas led to a virtual veto on any reference to mental states. There was also the influence of the causal-mechanistic position. In this type of explanatory approach, primate behaviour could be understood as acting under the invisible hand of natural selection without assuming individual agency (see Harré, 1984, for a critical discussion of this view).

We can certainly assume that monkeys and apes have intentions. The definition of intention that I follow here is that actions are intentional if they are directed towards the attainment of a future goal. Intentions are "mediators between desires and actions", so that from desiring something comes the intention to obtain it (Astington & Gopnik, 1991, p.45; Allen, 1995) and to seek the means to do so. In humans, studying intentions and distinguishing intentional actions from actions that fortuitously attain a result, is made much easier by asking questions like: "Why are you doing X?" (Frye, 1991, p.16). In studies of monkeys and apes, having intentions or, in the case of chimpanzees (*Pan troglodytes*), recognizing intentions in others, is a central issue in the interpretation of their social behaviour. The coordination of joint actions that is seen in the social tactics of primates often suggests intention to achieve a goal as the most convincing explanation. It is a crucial matter, when we come to examine social actions later, that the following questions be borne in mind: can these actions be understood as responses prompted by stimuli in the immediate situation, or are they the expression of flexible, intentional goal-seeking?

Usually intention means a capacity to control one's actions with some particular goal in mind (first-order intentionality) and more than this, to be able to think about the goals of others (second-order intentionality). In studies of deception, for example, in nonhuman primates, there are many indicators of first-order intentionality; there is also evidence that chimpanzees' mental capacities are considerably greater than those of monkeys. Studies have suggested that chimpanzees can be aware of others' intentions and so have a "theory of mind" (Premack, 1988; Premack & Premack, 1983; Premack & Woodruff, 1978). Whiten (1991, 1994) has suggested the term "mindreading" for the attribution of mind states to others, as in the following illustration involving the chimpanzee Sarah, the subject of Premack's experiments (Whiten & Perner, 1991, p.14):

(1) "Sarah *knows* that the bananas are in the red box" expresses a *first-order* mental state. In contrast,
(2) "Sarah *knows* the trainer *wants* to eat the bananas" expresses a *second-order* mental state and we have mindreading.

Chimpanzees and Theory of Mind

Discussion of the question "Does the chimpanzee use a theory of mind?" was initiated by David Premack and Guy Woodruff (1978). As they defined it, humans have a theory of mind when they make inferences about the mental states of themselves and of others: about desires, intentions, and beliefs. It is a "theory" because such states are only inferred, they are not directly observable and the inferences that are made about them can be used to predict others' actions. From the series of experiments that Premack and Woodruff carried out, they concluded that Sarah understood the immediate goals, wants or intentions of a human actor, a conclusion that has been accepted by many researchers (Whiten, 1993). Sarah was shown videotapes of the human actor confronted with a problem such as how to obtain bananas that were out of reach; the bananas were hanging from the ceiling and the solution involved using a box located in a corner (as Premack & Dasser, 1991, have since mentioned, Sarah had first to understand that there was a problem). She then had to complete the sequence when she was presented with photographs, only one of which in each trial showed the correct solution. She did this correctly in 21 out of 24 trials (Premack & Woodruff, 1978, p.516). It is true that there were points of uncertainty about the experimental controls and questions about how to interpret the chimpanzee's performance. Similar problems arose in the interpretation of these experimental results, as have arisen in interpreting so-called anecdotes from field studies. Here too experimenters have provided what is essentially an account of the chimpanzees' actions and then inferences are drawn about the mental process involved. In observational studies and experiments alike, the quantification of results leads to the enhancement of belief and consensus on interpretations.

Whiten (1993, p.374) summarized the outcome as follows: "the working conclusion at present is thus that the chimpanzee attributes at least *motivational states*, such as we label *want* or *intend*."

There was no firm evidence of chimpanzees' attributing beliefs to others (as contrasted to desires and intentions). Attribution of beliefs in children only develops at 4 years of age. The quality and nature of attributions at this age and onwards become much more complex. Whiten (1994, p.63) has proposed that this development in complexity can be understood as reflecting gradations of mindreading corresponding to the "depth of penetration that a mindreader makes into the mind of another." Following this line of argument, understanding that another individual's visual perspective is different from one's own, for which there is ample evidence in monkeys, is a relatively superficial perception. Understanding that others have wants and intentions is at a relatively more advanced level. Attributing beliefs to others is again more advanced and considered by many developmental psychologists as marking the full attainment of theory of mind in humans. Whiten (1994) goes on to conclude that theory of mind in children involves inferences about a variety of mental states and this knowledge is integrated and combined to make sense of, explain, and predict others' actions.

CHIMPANZEE INTELLIGENCE

A broad definition of intelligence would emphasize adaptability, but intelligence tests used for chimpanzees are often similar to classic tests for the *eduction of correlates* in traditional mental measurement. These tests assess the capacities to perceive objects as same or different, as belonging to overall categories, or as similar by analogy (Premack, 1976; Premack & Premack, 1983). Sarah, for example, learned associations between coloured plastic shapes and pieces of fruit, even if it sometimes took hundreds of trials before learning was achieved. There is further evidence of chimpanzees' capacity to understand *complementarity*, for example, between a bottle and a cup (Tanaka, 1996). Sarah could distinguish between same and different and she was trained to match objects—like to like. When presented with new objects, Sarah showed that she had learned the *matching rule*. Other chimpanzees, however, were not so successful and required much more training. Sarah could reason by analogy, understanding, for example, that half a banana is to a whole banana as half an apple is to a whole apple and indicating the correct answer much as a human would in the sentence completion tasks of conventional ability tests.

Using objects and plastic shapes as symbols for them, the Premacks suggested that chimpanzees could perform many tasks requiring them to represent an object mentally. Sarah's training in the use of words as symbols, itself a special attainment of chimpanzees, was evidently extremely important in these tasks. She had no difficulty in using the words "same" and "different". Sarah gave correct answers for "what is the colour of apple?" (answer: red) or "what is the shape of a cracker?" (answer: square) even though neither apple nor cracker were present. This kind of representational thinking, the Premacks argued, was a basic preliminary step towards the evolution of language. Mental representation has been defined, following a long tradition in cognitive psychology and neurobiology, by Delacour (1995, p.318) as "a state M of a system that maps another reality O." Cheney and Seyfarth (1990a, p.16) give their general definition of mental representations, for humans and nonhumans, as including images and "symbols that guide an organism's behaviour . . ." In the case of humans and other biological organisms, Perner (1991a, p.143) suggests that a definition of representation should affirm two elements: correspondence between the representation and the represented world and an interpretive system to make use of it as a "stand-in". Recently, Byrne (1997, p.298) has mentioned that great apes' accomplishments suggest that they "have a *representational* understanding of the world, one that can support simple computation. In everyday terms, they can *think*."

THE SELF CONCEPT: GREAT APES AND MONKEYS

What were the consequences of the Premack experiments for understanding chimpanzees' social behaviour? They opened up a far-reaching discussion of chimpanzees' intelligence and imputation of mental states to others. They also suggested the controversial notion that chimpanzees can separate a concept of the self from a concept of the other. Another line of research relevant to this possibility, pursued originally by Gordon Gallup and more recently by Daniel Povinelli, has tested whether chimpanzees and other primates can recognize their body image in a mirror. This has emerged as a capacity shared only by humans, chimpanzees, orangutans (*Pongo pygmaeus*), and possibly, gorillas (*Gorilla gorilla* species). Recognizing the body image in a mirror might also imply a capacity to think reflectively, to be self-aware (Gallup, 1970, 1982), and there has been considerable discussion around this issue (see Parker, Mitchell, & Boccia, 1994).

Gallup's procedure was as follows: he tried showing pre-adolescent chimpanzees their mirror-image. Later, he anaesthetized the chimpanzees, painted red marks on eyebrows and ears and, when they recovered, presented the mirror. The chimpanzees now touched the marked areas while watching their reflection. In orangutans, similar results were reported, with recognition appearing after about three days of mirror trials. A variety of monkey species were tested without success and showed no sign of recognition, even after very prolonged familiarity with mirrors of up to a year or more. Nearly 20 species have been tested and have shown little evidence of this capacity (Gallup, 1982, 1983; Gallup & Povinelli, 1993; Mitchell, 1993; Platt & Thompson, 1985). For gorillas, there has been a positive result only in one case (Patterson & Cohn, 1994; Tomasello & Call, 1997). From evidence on anatomy, protein structure, and

DNA, gorillas could be expected to be nearest in mental capacities to the other great apes and to humans.

The gap between monkeys and apes in their perceptions of mirror images is well confirmed (Gallup & Suarez, 1991). Monkeys see a mirror image as another individual, a possible rival. A mirror therefore may incite aggressive actions towards others. After the introduction of a mirror to four Japanese macaques in a zoo study over several months, the number of aggressive signals in the group increased, especially from the alpha male (Neel, 1996).

Only during the second year of age does a child come to realize that a mirror-image is his/her own, rather than that of another child. In humans, does recognizing one's own body in a mirror indicate awareness of self? This does seem to be the case because the child has to be able to understand the distinctiveness of two mental constructs, one of itself and one of its own reflection in the mirror (Perner, 1991b). Could a less demanding explanation account for chimpanzees' self-examination in front of mirrors? Might it be no different from the self-grooming that occurs without mirrors? (Heyes, 1994). This criticism has been confronted by a further study, returning to the question of whether chimpanzees used mirrors as a source of information about their bodies (Eddy, Gallup, & Povinelli, 1996). The interesting finer points of this experiment involved comparing 11 older (7–10-year-old) chimpanzees with 12 younger (3-year-old) ones. The comparison of self-grooming rates in front of mirrors with the frequency without mirrors, showed that the older chimpanzees did make self-exploratory actions in front of the mirror which otherwise they did not do: the self-exploratory behaviour consisted of inspecting parts of the body not directly visible, such as teeth and the inside of the mouth (see Povinelli & Preuss, 1995, for illustrations of these; Povinelli, Rulf, Landau, & Bierschwale, 1993).

There are many questions still to be answered about the mirror experiments, if only because, in humans, the self is a social construct. It is constructed from the "mirror reflection" of others' actions and attitudes towards the individual during infancy and childhood. There is interesting work by Meltzoff (1990) on parental imitation of human infants, acting as a "social mirror". Parker and Milbrath (1994) have reviewed recent studies that highlight the self-construction that develops through the imitation and role-playing of children's play. The importance of the social process in the development of the self was well recognized in the work of George Herbert Mead (1934). He suggested the duality of the self—the *I* as a spontaneous unselfconsious biological development and the *me* which is built as a product of social interactions. The *me* makes self-awareness possible in relation to a network of other selves. Future considerations of notions of the self in nonhuman primates should take into account this line of approach. One interesting lead in research was that chimpanzees raised in isolation could not recognize themselves in a mirror (Gallup, 1977). It may be that chimpanzees' social interactions have helped them form a view of the self in which body image is a conspicuous aspect, the first step towards awareness of a *me*.

William James (1892/1961) had suggested that humans have an image of the self (self-concept) and also understand this image as object (self-awarenesss), in Mead's words (1934, p.136): "the self has the characteristic that it is an object to itself." According to Hart and Fegley (1994), mirror self-recognition in children in their second year shows that the child is capable of self-awareness. Yet the emergence of this awareness is only one of several dramatic changes at this age which notably involve symbolic play and, in social behaviour, reciprocal cooperative actions (Kagan, 1981). The clear indications of studies of apes' recognition of their body-image in mirrors encourage us to speculate also about their achievement of understanding, in the social sphere, of the reciprocal demands of relationships with other individuals (as will be discussed in later chapters). Here, again, is an area where chimpanzees come closest to humans in their achievements. A pattern of related and distinctive cognitive abilities now seems to be emerging, as a result of recent studies.

Tentative acceptance that, in these mirror experiments, chimpanzees show an awareness of self, would be important in taking us towards an understanding of their social relationships. Awareness of self might be yet another expression of chimpanzees' "limited aspects of theory of mind" (Povinelli & Preuss, 1995, p.423). In humans, awareness of self involves awareness, in interaction with others, of *their* distinctive selves. For Mead (1934, p.141), the exercise of social intelligence depended on individuals being able "to take the roles of, or 'put (themselves) in the place of' the other individuals implicated . . . in given social situations." In recent studies, as we have seen, there is cumulative support for chimpanzees' abilities in thinking ("theorizing") about the mind states of others and, following Gallup's (1982) hypothesis, knowledge of self is closely connected to these abilities.

PIAGETIAN SENSORIMOTOR INTELLIGENCE

The progressive sequence of stages that Jean Piaget (1963) proposed to explain the cognitive development of the child, has, in recent years, received some modifications as a result of subsequent empirical testing with varied experimental tasks. Nevertheless, as a general framework, Piaget's stage sequence is proving useful as a means of systematizing and testing our knowledge of monkeys' and apes' abilities (see, for example, Russon & Bard, 1996; Parker & Russon, 1996). As an illustration of this interdisciplinary transfer we can consider the six sensorimotor substages that Piaget proposed for the first two years of a child's cognitive development. In the final stage (6) the child was able to represent objects and future actions mentally, understanding, for instance, that a hidden object continues to exist even when it cannot be seen (*object permanence*), inferring causes from effects, and foreseeing effects, deriving from a cause. Several studies have shown that chimpanzees as well as orangutans and gorillas, can attain the achievements of this stage. Stage 6 competence is but one of the many cognitive talents of chimpanzees (Flavell, 1993). Sensorimotor Stage 3 (the goal of behaviour only emerges in the course of a sequence of actions) possibly fits many monkey social episodes but it should be worth searching for more evidence, using the stages as a guide for inferring particular forms of thinking. Researchers thus have a useful strategy at their disposal.

Apes' abilities appear to be well beyond this period of sensorimotor development, placing them at more advanced levels. It can be argued that chimpanzees, in using *words* as *symbols*, reach a level analogous to the Piagetian early preoperational stage in the mental development of 2–7-year-old children, a stage characterized by symbolic thinking. In a long-term study, Miles (1994) has demonstrated the equivalent attainments of the male orangutan, Chantek, in using rudimentary symbolic categories in word form, naming objects that were not present. Current research is tentatively exploring the competence of orangutans in famous Piagetian-type tests for the concept of *conservation*: in this case, the ability to understand that quantities of liquids do not change despite being placed in containers of different shapes (Call & Rachat, 1996). Orangutans have not been shown to achieve conservation in its full Piagetian sense as yet.

Attainment of competence analogous to Stage 6 was demonstrated in experiments with chimpanzees and gorillas during the 1970s and 1980s (Antinucci, 1989) and recently again for chimpanzees and orangutans (Chalmeau & Gallo, 1996; Chalmeau et al., 1997). There have been no equivalent sure demonstrations for monkeys using Piaget's test for mentally representing an object when it is out of view (the object permanence test). In this type of experiment, researchers have been careful to limit the number of trials, thus avoiding learning effects. In an experiment with rhesus monkeys, de Blois and Novak (1994) thus created conditions for spontaneous search to retrieve the hidden (displaced) object. The monkeys' attainments were typical of Stage 5—that is, they found the hidden object provided they saw it being hidden (visible displacement). In tests where they did not see it hidden but had to infer that it had been (invisible displacement: Stage 6), they were not successful.

In general and thus far, the gap between monkeys' and apes' performance has been suggested in a number of contexts including chimpanzees' attribution of states of mind to others (Povinelli, Nelson, & Boysen, 1990) whereas monkeys do not make such attributions (Cheney & Seyfarth, 1990a; Povinelli, 1994). Chimpanzees appear to achieve well beyond the Piagetian sensorimotor stages of an infant's first two years, where most (but not all) monkeys' capacities appear to centre, although experiments continue to test monkeys' abilities in an innovative fashion (Smith et al., 1997). There are persistent efforts in recent research to present information relevant to the apparent gap between monkeys' and apes' performance. In computerized tracking tasks, rhesus monkeys were evidently able to follow mentally the movements of a target when it passed out of sight (Filion, Washburn, & Gulledge, 1996), in an experiment pointing to future research directions.

Cautions about Piagetian Tests

Some words of caution need to be mentioned. First, when a task is used as a test of a certain level (stage) of cognitive attainment, it has to be remembered that in Piaget's model of development, many aspects of information-processing skills are involved at any one stage (Etienne, 1989). Second, Piaget's model was intended to demonstrate the sequence of human development in infancy and childhood. Using it to try to systematize the cognitive achievement of adult monkeys and apes involves tasks that are similar, to some extent, but the Piagetian tests, from the infant's first year onwards, rely very much on verbal instructions. Third, the resulting achievements or failures on Piagetian-type tests can only suggest inferences about monkeys' and apes' abilities that are merely analogous (to an uncertain degree) to human abilities. A fourth problem derives from the fact that Piaget's stages relate mainly to instrumental activities (manipulating objects) rather than to social activities. It is in the social domain that, in their natural habitat, monkeys and apes really show their abilities. Having said this, the standard Piagetian incremental stages do provide invaluable guidelines for discussing the cognitive achievement of any one particular species (never monkeys and apes in general). Each succeeding stage is constructed on the basis of abilities attained at preceding levels and each stage represents the highest level of abilities at a given point of development (for discussion of these advantages, see Russon & Bard, 1996).

Piagetian experiments up to now have been tried with only a few species in testing for achievements in representational thinking: notably with chimpanzees, gorillas, and orangutans, and with several monkey species. We have to await further studies for more information on comparisons between species. Without verbal instructions the nonhuman versions of the tests have inevitable limitations. The attentional demands of the tests often produce difficulties and nonhuman primates also lack the support of other kinds of knowledge that the developing child has acquired through language.

IMITATION AND OTHER ABILITIES

Studies of apes' imitation learning have been limited because the motivation to imitate, in the context of experiments, has often been absent (Russon & Galdikas, 1995). The ability to imitate appears to be related to visual perspective-taking and to self-awareness. Summarizing the existing evidence, Parker and Milbrath (1994, p.122) argue that

> the imitative capacity of great apes allows them: . . . vicariously to experience the subjective state of other animals at least at the level of enacting their behaviours . . . in playfully reversing a learned role, for example, by playing teacher. . . . These abilities may allow great apes to anticipate the reactions of other animals, to plan strategies, to deceive them or to adopt complementary roles.

Some researchers have made useful contributions by considering mirror self-recognition *and* imitation together.

Hart and Fegley (1994), for example, argue that if visual self-recognition in children implies a self-concept then there should also be evidence of (social) activities revealing a mental model of the self, such as imitation and deception. These authors imaginatively refer back to Baldwin's (1902) view that the development of self-awareness and the self-concept are *products* of *imitation*. Imitation appears to have several aspects: its cognitive qualities involving the sheer degree of competence that enables imitation of complex actions and furthermore, the *imitation process* that, as it occurs, changes and extends the self and its abilities. Thus, a child imitating anothers' actions comes into contact with the mind and intentions of the other and is encouraged towards "construction of a theory of mind" (Hart & Fegley, 1994, p.155). We can apply this particular argument, by analogy, to apes. Apes both imitate and deceive, often in spectacular fashion (see later) but monkeys also can deceive although they do not imitate. However, Custance and Bard (1994), in an experiment with chimpanzees, presented counter-evidence to the view that imitation is necessary before self-recognition can develop. Whiten (1994, p.65), discussing apes, argues that imitation requires "a transposition of visual perspective" in translating the actions of the individual being imitated into the equivalent movements of the imitator. He also suggests that imitation may involve some understanding of the other individual's intentions. This ability is receiving much current attention because of the theory of mind debate. Russon (1997, p.178) mentions that "imitation is considered to reflect symbolic mental processes because it generates behaviour from mental images created by observation, without direct experience." Experimental tests here again suggest a cognitive gap between monkeys and apes (Mitchell & Anderson, 1993; Whiten, 1996a).

THE COGNITIVE GAP BETWEEN MONKEYS AND APES

In fairness, it has to be said that there is often disparity between the performance of monkeys in laboratory tests and their activities in the wild. Different habitats and social patterns have favoured distinct abilities. Monkeys do better in activities that are similar to those shown in their natural environments—they each have their own style. Squirrel monkeys (*Saimirii* species), for example, have the same bouncy style of approaching problems as they use in the forest (Jolly, 1985); titi monkeys (*Callicebus* species) are more cautious; in their natural habitat they keep under cover, hiding from predators and moving slowly (Fragaszy, 1981). Monkey intelligence tests have tended to be somewhat arbitrary, using sound tones or lights unrelated to the natural environment (Cheney & Seyfarth, 1990a). In a monkey's typical day, in the forest or savanna, mental abilities are exercised continually in acting towards other individuals. Much of the time, in the wild, monkeys direct their attention not to objects, but to social events. The intelligence of monkeys and apes is above all *social* intelligence. This is a well known view in discussions of primate intelligence (Humphrey, 1976, 1988; Jolly, 1966, 1988).

In the wild, monkeys and apes need to know how to find food, how to avoid predators, and how to act towards a variety of individuals of different ages, gender, and dominance. Their social community provides them with opportunities for learning how to do these things. Social skills are a prime requirement. Monkeys and apes need to assess the motives and future actions of others, to judge what tactics to follow and when to change them, even to use some degree of forward planning (Humphrey, 1988).

The young, in primates, have a very prolonged period of dependence, which is longest in humans, followed by apes and monkeys. During this period (several years in chimpanzees) the young are in contact with experienced elders and have many opportunities for social play, trying out future adult stratagems (Smith, 1984). Young monkeys are dependent on the troop for protection and training. In avoiding predators, for instance, they take their cues from others in the troop in a process of social learning (Jolly, 1966, 1988). Eventually, in the evolutionary timescale, the mental abilities that were developed to manage social relationships, found expression in the cultural creativity of humans (Humphrey, 1988).

There have been intensive research efforts to throw more light on monkeys' abilities. It has been one of Cheney and Seyfarth's (1990a, 1990b, 1991) important contributions that they placed at centre stage the exploration of monkeys' thinking *in* and *about* social relationships. These researchers, for instance, tested the ability of rhesus and Japanese macaques to attribute knowledge or lack of knowledge to others. One test required a mother to vocalize a signal about the presence of food to an infant who had not seen where it was hidden, although the mother had. There was no increase in the mothers' calls in this condition as compared to the control condition in which infants had seen where the food was placed. Discussing this and other experiments with monkeys, Whiten (1993, p.381) concluded that "only chimpanzees read the intentions and perhaps the knowledge of others"; monkeys have not been shown to do so.

A valuable approach to monkeys' social abilities was provided by Verena Dasser's (1988a, b) experiments. She presented, to two adult female monkeys, a series of photographs of pairs of other individuals belonging to a large captive group of Java monkeys at the University of Zurich. Some of the photographs were of mothers and offspring at a variety of ages. The females had to be able to pick out these pairs and did so successfully in a convincing series of trials. Subsequently they performed similarly well in choosing sibling pairs. So far as possibilities of confounding are concerned, in these results, it is not known if continuous social interactions in this group of monkeys, over a period of years, provided opportunity for the formation of social categories based on temporal and spacial proximity rather than on awareness of kinship, as Thompson (1995) has suggested. Such possibilities would exist in any captivity group of long-standing association. Menzel (1997) makes a further comment about this study when he states that the precise basis for the choices is not known currently.

So far as the comparisons can be made, despite problems in interpreting the experimental results, there is agreement, among some researchers, on the relatively advanced levels of cognitive attainment in chimpanzees (Whiten & Byrne, 1997; Whiten, 1994). It has been confirmed many times that chimpanzees perform better than monkey species in a range of experimental tests but the issue of a cognitive gap is a lively area of controversy (Visalberghi & Fragaszy, 1990; Whiten, 1996a). Some researchers deny that such a gap exists, because strictly comparable tests have not been applied to monkey species and, in any case, the results at present are unclear (Tomasello, 1996). Contemporary researchers are using the Piagetian stages of cognitive development in children as a standard framework for comparing primate species (Russon & Bard, 1996). From the Premacks' series of experiments as well as from other researchers' work (Rumbaugh, 1990), it seems clear that chimpanzees are well ahead of other species. Experiments have shown that a chimpanzee is able to supply the missing element in a causal sequence of a trainer's actions, revealing understanding of cause-and-effect: the capacity for causal reasoning (Premack & Premack, 1994). Although it is not within our main area of discussion here, tool use by chimpanzees has also provided evidence of this capacity, (Boesch, 1991; see also Inoue-Nakamura & Matsuzawa, 1997, for a discussion of the social learning process in tool use). There are no equivalent indications for monkeys (Visalberghi & Limongelli, 1994).

SOCIAL INTELLIGENCE

Traditional Piagetian tests are not likely to help us much at the social psychological level and we have to look elsewhere for evidence of social intelligence, which I define as autonomous, flexible behaviour leading to the attainment of social goals and reflecting capacities for choosing alternative tactics in relation to others (for discussion of a variety of definitions, see Whiten & Byrne, 1997; Strum, Forster, & Hutchins, 1997). Monkeys are profoundly sociable and adept social actors. Do monkeys understand cause-and-effect (do they use causal reasoning) in their social tactics? When an individual acts against another while appealing to a third for help, is this only the result of association learning from previous experiences? These questions about causal reasoning need to be taken into the social context.

It is difficult to find incidents that are explained only by inferring understanding of cause-and-effect. Adult female chacma baboons (*Papio cynecephalus ursinus*) use close-contact grunts in friendly approaches to an individual lower in dominance rank, and subordinates often make fear-barks in reply. In a series of playback experiments, grunts and fear-barks were played in reverse order:

fear-barks first followed by the close-contact grunts (Cheney, Seyfarth, & Silk, 1995b). It was hypothesized that listeners would look for longer periods towards reverse-order sequences than they did in the baseline periods established for correct-sequence sounds. This did occur, suggesting that the baboons' attention to the reverse-order sequences signified understanding of a cause-and-effect sequence between grunts and fear-barks.

This field experiment is a useful step in studying causal reasoning in the social domain. As usual, in playback experiments it is also a question of how much to infer from the reported durations of looking by the listeners. The weight of explanation rests rather heavily on durations of visual response. In this case, causal reasoning could be inferred but we cannot be absolutely sure. At the same time, it should be said that experiments such as this, and the work of Andrew Whiten and Richard Byrne in their emphasis on social relationships, are crucially important because so much of apes' and monkeys' cognitive capacities have their natural and most frequent expressions in social situations.

DECEPTION AND MINDREADING

Next, we consider studies of a frequent expression of complex thinking in primates—competence in deceiving others. Reports have been reviewed of deceptive actions and intentions to deceive, across a large number of monkey and ape species, including experimental and field studies (Byrne & Whiten, 1985, 1987, 1988; Whiten & Byrne, 1988a, b). The question asked was: does an individual anticipate the intentions of others in such a way as to deceive them? This is illustrated, for example, in the case of a fleeing juvenile baboon chased by several adults grunting aggressively (Byrne & Whiten, 1985, p.671). The juvenile suddenly stops, "stands on hind legs and stares into the distance . . . (the pursuers) stop and look in this direction; they do not threaten or attack." Did the juvenile intend to deceive or had he learned the trick by imitating others? Did the deception involve thinking about others' intentions? Or was it no more than repeating an action seen and learned on previous occasions? These were some of the interesting questions that the incident raised.

A questionnaire was sent to 115 primatologists. From the questionnaire responses, 75 descriptions were taken that qualified as tactical deception, where tactical refers to flexible deceptive actions as distinct from fixed displays of body camouflage or mimicry. Some notable examples of deception were assembled, not only in chimpanzees, but also in baboons and other monkeys (Whiten & Byrne, 1988b).

One was of a female hamadryas baboon (*Papio hamadryas*) grooming a subadult male behind a rock concealing him from a dominant male. At first in full view of her unit leader, the female spent 20 minutes moving gradually until only the lower part of her body was hidden behind a rock. Also hidden by the rock was another male; the female now groomed him with her concealed hands

(Kummer, 1968). Her behaviour showed that she was able to appreciate the other individual's visual perspective, an important cognitive attainment. In this situation, the time taken by the female to position herself showed that she was aware of the male's visual perspective as different from hers (Byrne & Whiten, 1987).

What can we infer from this about the social intelligence of baboons? Did the baboon think that if she positioned herself in that way the male would not see her; did she intend to deceive? The answer must be positive. But this example of visual perspective taking is *only* that, and is at a lower level of cognition, in the Piagetian sense, from chimpanzees' awareness of others' intentions (Whiten, 1996a). Whiten (1993, p.381) suggested that monkeys' ability to take another's visual perspective is "only shallow mindreading", a kind of mindreading at a more rudimentary level than that of chimpanzees. As mentioned earlier, this argument contends that understanding differences in *seeing* demands less penetration into the mind state of another than understanding the *intentions* of another.

Understanding that the viewpoint of others is different to one's own view was, in the classic Piagetian experiments, achieved by children towards the end of the pre-operational stage at 7–8 years of age (Piaget & Inhelder, 1956). Subsequently, researchers have estimated that children at 3–4 years of age (and in some simple tasks, even younger) can achieve appreciation of others' perspectives. In one experiment a doll is placed in an apparatus with partitions similar to walls. The 3–4-year-olds show that they understand its visibility for another doll at different locations relative to the walls (Donaldson, 1978). This task seems similar to the baboons' perspective taking.

Using the terminology of child development studies, it seems plausible that the female baboon did not act *procedurally*; she did not just do the action. On the contrary, she behaved in a *declarative* way, making a causal inference about her positioning and the other's view of that position. She must have adjusted her position to the male's field of vision, to a mental image of the male's perspective. Procedural *knowing how* corresponds to behaviour following a particular stimulus, it is a reaction without thinking. Declarative *knowing* implies thinking about the situation and generalizing from one situation to another, so that actions are guided by changing circumstances (Cheney & Seyfarth, 1990a; Whiten & Byrne, 1988a). This incident of the baboon's concealment, although it was merely an anecdote, a field observation, did open a useful discussion of monkeys' abilities.

Byrne and Whiten's List of Deceptions

Deception involves relating to another individual's intentions and trying to influence them by sending a false message. A total of 13 categories for different forms of primate deception have been suggested (Whiten & Byrne, 1988b).

TABLE 2.1
Type of deception

Species	*Conceal (A)*	*Distract (B)*	*Create Image (C)*	*Manipulation in Triads (D)*	*# of Primatologists Reporting Incidents*
Baboons and Vervet Monkeys	X	X	X	X	59
Chimpanzees	X	X	X		10
Gorillas	X	X			12

Adapted from Whiten & Byrne, 1988b.

Some of the most important ones are now illustrated from the list compiled by these authors, as follows: Conceal (A), Distract (B), Creating an image (C), and Manipulation (D). Baboon descriptions were much more numerous than others, but this was partly due to the larger number of researchers working with them.

Table 2.1 shows an abbreviated version of the major categories. Chimpanzees were reported as concealing (A), distracting (B), and creating an image (C); altogether nine classes of deception (A1, A2, A3, etc.). No chimpanzees were reported using the stratagems under category D involving third parties, manipulating one individual to act against another. This may have been because chimpanzees were too intelligent; they could see through such ruses. The details of every category are now given, as follows:

A. Concealment

A1. Concealment of something: hiding food or hiding part of the body, as in the female baboon's concealment of herself behind a rock.
A2. Acoustic concealment: suppressing vocalizations, as has been observed in chimpanzees, baboons, and howler monkeys, to avoid aggression or gain advantage over others.
A3. Avoidance of looking at or attending to something or somebody.

B. Distraction

B1. Distracting by looking away towards another direction to shift another's attention: as in distracting a pursuer by stopping and staring into the distance.
B2. Distracting by looking away, with vocal signal. A vervet monkey giving predator alarm calls to distract a rival is an example of this (discussed by Cheney & Seyfarth, 1990a).
B3. Distracting by leading away: from a food hiding place, returning to it later alone (Goodall, 1972, p.96).
B4. Distracting by grooming the fur, embracing, presenting: examples are found in chimpanzees, rhesus monkeys, baboons, and other monkeys. An example is a

female baboon's grooming of a feeding male until he relaxes and she snatches the food.

C. Creating an Image

C1. Creating a neutral image: by nonthreatening behaviour. Aversion of gaze, as described in the initial contacts between a female and male baboon before forming a consortship, is a good example of this. Another example of creating an image is a description of a male chimpanzee avoiding showing his fear-grin to a rival by using his fingers to push his lips back over his teeth, with his back turned to his rival. He did this no less than three times before the grin disappeared and he turned around to face the rival (de Waal, 1986a). Had the chimpanzee merely learned that if he grimaced the other would win?

C2. Creating an affiliative image: as when a female monkey grooms the mother of an infant, so that she can reach and touch the infant.

D. Manipulation in a Triad

An example of D category occurs in chacma baboons (Byrne & Whiten, 1987). A juvenile screams in distress when near a feeding female and the juvenile's mother chases the female away, thus leaving the food for the juvenile. This incident was described as character assassination. The juvenile Paul screamed loudly, which baboons do not do normally unless threatened. Paul was seen playing the same trick on different victims on other days. Had Paul accidentally produced this effect, gaining the food, on the first occasion and did he now repeat the behaviour because of conditioning? The tactic was used only when the mother was out of sight but within earshot. It may have been selective reinforcement to these special conditions, but the intention to deceive and anticipation of what ensued between the two adults could not be ruled out.

CONCLUSIONS

In this chapter, aspects of intentional actions have been discussed and examples given of primates acting to their immediate and, possibly, longer-term advantage in what seems to be a thought-controlled fashion. Research has shown clear indications that apes have a concept of the self; that chimpanzees understand that others have intentions. Information now becoming available on monkeys' and apes' deceptive actions provides the beginnings of an inventory of mental powers expressed socially, as we have seen in this chapter. Current field research is moving studies of monkeys' and apes' cognitive abilities into fascinating areas of their social competence. In addition, Piagetian-type experiments add yet another source of swiftly accumulating research reports.

Implicit in the interpretations of all these studies is the use of human parallels as points of reference for making any systematic appraisal. The parallels are used explicitly in cases where thinking in nonhuman primates has been compared to children's pre-linguistic and early development. Such comparisons justifiably occur in efforts to clarify questions about monkeys' and apes' social actions. In the next chapters I will continue to focus on the relationships, tactics, and flexible social skills of primates.

CHAPTER THREE

Primate Communication and Social Psychology

COMMUNICATION

Starting with a view of communication from zoology, in this chapter we will consider questions about the interpretation of primate signals by observers. The task of interpreting the meaning of messages sent in social interactions will be illustrated by a case study of "notifying"; a sequence of signals between male hamadryas baboons. We will see how researchers' interpretions of these sequences of communication have developed as the analysis has deepened over several studies and there has been a change from an interest in the tactics of the foraging march (the baboons' extensive and daily search for food) to an emphasis on the negotiation and control of social relationships.

To understand more about monkeys and apes, a close examination is needed of their social actions and relationships. *Social actions* involve signals sent from one individual to another in any modality; auditory, visual, or olfactory (as in scent-marking). When these actions occur in joint sequences, between individuals, they can be called *social interactions* which, over the longer term, may stabilize into *social relationships*, such as female/male consortships or alliances between pairs of females or males (Hinde, 1982). The first and central topic of *social actions* often comes under the heading of *communication*, the use mainly of visual and vocal signals. I do not want to enter into the morass, as described appropriately by Bruner (1975, p.261), that awaits anyone trying to establish whether a signal is really intended. Bruner's solution was to observe the context and to scrutinize the relationships between signaller and recipient: "demander and complier, seeker and finder, task-initiator and accomplice, actor and prohibitor" and this is the investigative approach to be followed here.

All sensory channels are used by primates in communicating. We will need to examine every aspect of inter-individual communication if we are to understand the social psychology of monkeys and apes. A view of social actions or communication as "influencing or being influenced by, the behaviour of other animals" is found in Marian Dawkins' *Unravelling animal behaviour* (1995). She mentions (p.100) that there is "a continuum with small inconspicuous and highly economical signals at one end and overt physical violence at the other." In this book, the "small, inconspicuous signal" will especially be the subject of study.

In zoology, communication is divided into four aspects: signal, motivation, meaning, and function (Smith, 1965, 1977). The first three are familiar concepts in social psychology: *signals* or messages occur in the form of vocal sounds, body movements, facial expressions and so on (even scent-marking in some species could be included). *Motivations* can be inferred from such actions: motivation to achieve a social goal. The ensuing reaction of the recipient of a signal leads us to infer *meaning*, while *function* points to the evolutionary advantage or consequences of the signal. The first three concepts are all proximate in that they can be used to explain what happens in social relationships in the immediate context of events. The fourth concept, *function*, however, is not familiar at all in social psychological studies and is often treated with suspicion because it might reduce explanation to underlying biological drives. In fact, what the concept of *function* does is to introduce the idea that social behaviour is influenced by the operation of natural selection on individuals.

The relevance of evolutionary theory to human behaviour is often obscured by the vast elaborations of social, cultural, ethical, and political diversity. The discipline of social psychology itself provides no master theory and no overall logical framework such as evolutionary theory, to guide research programmes. Social psychology only has proximate theories; theories that attempt to explain thinking and its expression through actions, in the immediate social context. But, in zoology, proximate theory can contribute to evolutionary theory because functional explanations can be strengthened if we have more knowledge about proximate social process (Dunbar, 1988).

A SOCIAL PSYCHOLOGICAL PERSPECTIVE

A social psychological approach aims to extend knowledge of motivation and the meaning of actions. There are many accounts of social interaction in primate studies, as the number of studies cited in these chapters confirms, but even so it is relatively rare to find published reports of social sequences that are presented in close detail, signal by signal. Probably such a reportage in full detail is considered simply too cumbersome to publish. Perhaps there are also other reasons. First, some researchers may have been mostly interested in biological and ecological aspects of their subject matter, considering social aspects as anecdotal rather than as a central topic for research. Second, it is difficult to track many species in the wild, especially arboreal species. Most of our detailed knowledge of social interaction comes from ground-living or partly ground-living species such as chimpanzees, gorillas, baboons, and macaques. Third, researchers have often selected characteristic incidents, rather than giving actual descriptions. Occasionally, a hindsight narrative is provided, as in Shirley Strum's (1987) book *Almost human*. Here the interpretation skips referring to social actions; for example, in discussing dominance contests between males (p.77):

> Ray and Big Sam were a good example. One day Ray would come out ahead, the next, sometimes even within the hour, Big Sam would be the winner. Undaunted, Ray would try again until he won. Stubbornly, Big Sam would refuse to let the issue drop.

These were interesting interpretations but what did the two males *actually* do? Baboons are "put in their place", are "wary", "secretive", "work out their relations" with others, have "degrees of attachment" and so on, but there is no clue to the actions from which these inferences are drawn. Kelly Stewart's (1988, p.14) reference to this style of reporting as "little scenarios" in her review of Strum's book was appropriate, as was her send-up of the style:

> Theo's a bitch and her mother is socially brilliant. And Ray's putting the moves on Peggy while Sumner's trying to break it up.

The style marred Strum's extremely important theme, which was essentially to give accounts of baboons as intelligent social actors.

Anecdoctal references to *typical* behaviour, such as "a chimpanzee usually greets another thus", have often been the rule. Researchers have asked us to trust one illustrative description as sufficient. From this attenuated form of reportage it is hardly possible to develop a social psychology of nonhuman primates. Fortunately, a growing number of contemporary researchers do give detailed descriptions and their material will be discussed throughout this and the following chapters.

SOCIAL PSYCHOLOGICAL TERMINOLOGY

Since the 1980s a social psychological approach has influenced the terminology used to explain the actions of monkeys and apes. A notable example was Robert Hinde's (1983) use of the term "social approval" for the feelings that monkeys probably express when they groom others or when they make the *presenting* gesture (turning the rear towards another). Both these actions are friendly and occur between males, between females, or between females and males in nonsexual friendships as well as in courtship and mating behaviour. There will be more to say about them in later chapters because they occur very frequently in many different species.

Does the use of terms such as social approval or friendship introduce anthropomorphic error, by attributing human sentiments to nonhumans? In answer to this: there is always some degree of anthropomorphism in descriptions of social actions (Asquith, 1981, 1984). This seems to be the case whatever the terminology. Descriptions in ordinary language inevitably have the characteristic of *generic anthropomorphism*. They have this characteristic because language was intended for human social behaviour and will always carry some of the meanings appropriate for the human context. Terms such as threaten, appease, and greet are associated with human purposeful actions. Inevitably, accounts of social activities tend to assume purpose in the actors. To that extent a generic anthropomorphism is unavoidable.

Ethologists have been unsuccessful in finding descriptive terms that avoid imputing purpose to monkeys. Their efforts result in a mixed vocabulary where some words suggest free agents with purposes and others refer to monkeys as automata. Hans Kummer (1984) defended using the vocabulary of the behaviourist laboratory—such words as "respond" and "emit", while at the same time mixing in words such as "prefer" and "decide". He argued that the latter were only used for what had so far escaped a satisfactory explanation but it is doubtful if his argument entirely carries conviction, even though this experienced researcher's opinion merits respect.

Apart from specialized behaviour such as tool use in chimpanzees or foraging strategies in baboons, primate mental capacities are revealed mainly in complex social actions. The social behaviour of monkeys and apes is distinctive but, in fact, describing and interpreting it involves using humans as a source model (Harré, 1984). The same words are used for describing both human and nonhuman social actions. Language has evolved in and about human societies; every descriptive term we use refers primarily to human actions and using them for monkeys' or apes' social behaviour tempts us to compare them with human ones. This inevitable degree of association is not the same as attributing human motives to nonhumans, which can justifiably be said to be anthropomorphic but we should still be alert to the problem. When nouns like aggression, cooperation, competition, courtship, friendship, alliances, reconciliation, and conversation are used, it has to be noted that they refer to behaviour that is distinctive to each species. A critical approach has to be maintained because we just cannot escape the effects of using terms that ordinarily refer to human behaviour.

It is also impossible to avoid referring to experiences similar to those of humans in our explanations of what monkeys or apes do, because of their phylogenetic relatedness and physical similarity (Fisher, 1996; Mitchell & Hamm, 1997). In an experiment with chimpanzees at the Yerkes laboratories in Georgia, a determined effort was made to ban anthropomorphic description. All that resulted was a report of specific acts in which no order or meaning could be found (Hebb, 1946). True, there could be major errors through assuming too much similarity to humans. One safeguard against anthropomorphic error is the obvious one of parsimony, by applying Lloyd Morgan's canon and avoiding more elaborate explanations if a simpler one will do (see Burghardt, 1985). This has been a classic antidote but it need not follow that a simpler explanation is more adequate. More adventurousness has been advocated in trying to explain nonhuman primate behaviour (Dunbar, 1988; Hinde, 1983). Social psychological terms and theories may help to uncover more of the subtle social actions of monkeys and apes than are known currently.

MEANINGS

An observer will assign a meaning to a monkey's signal *if* the interpretation is confirmed by the subsequent behaviour of the monkey and the recipient of the signal. Comparisons with human actions inevitably underly our interpretations and we should "try to make the comparisons sharper" (Harris, 1984, p.134). Social psychologists have wrestled with problems that arise for observers in attributing meanings. For philosophers, the *other minds* problem raises extremely difficult issues (Jamieson & Bekoff, 1996). The basic problem of ever being able to know what others are thinking, is immensely greater when it comes to second-guessing what might be the intentions of nonhumans.

In the next few pages I discuss a search for the meanings of social actions by considering the greeting signals of male hamadryas baboons. Questions about these signals were not settled even after research over long periods by field workers. They tried to assign meanings that could be agreed on, and changed their interpretations as fresh observations were made.

NOTIFYING

What was its Meaning?

In the early studies of the social behaviour of hamadryas baboons, ritualized greetings between males, with presenting and touching, were thought to be important for group cohesion during the foraging march. The greeting behaviour of leading males, directed to other males in the troop, was understood to notify them of changes in direction. This was a plausible functional explanation, recognizing the importance of the greetings for the structural organization of the hamadryas troop (Kummer, 1967).

Over a decade later, the interpretation was modified in the light of more research by Abegglen (1984), giving new indications about the adult males' relationships with other males. It was found that males notified selectively certain others. The males receiving the signal were either bachelors (mature males with no females) or had only recently formed a tie with one female and were on their way to recruiting more female partners. The explanation at this point was changed, now referring more directly to the nature of the social relationships

TABLE 3.1
Notifying Between Males

	Receivers			
Notifiers	*Unit Leaders*	*Initial Leaders*	*Bachelors*	*Total*
Unit leaders	13	136	6	155
Initial leaders	26	156	6	188
Bachelors	10	27	23	60
Total	49	319	35	403

Adapted from Abegglen, 1984.

between the males. So, the greeting was done to avoid attack from rivals, to disarm competition during the foraging march. The meaning of notifying to the recipient of the greeting was not only to inform of the direction of march but was also a notification of friendly intent. This interpretation was supported by observations of hundreds of incidents of notifying between the males in the troop. It was more adequate because it was an explanation that was comprehensive of both the structure of the group and the relationships of individuals.

Jean-Jacques Abegglen whose doctoral study was supervised by Kummer, took up the interpretation of notifying, specifically. He contended that "a more precise aspect" of notifying had emerged from experiments by Kummer, Goetz, and Angst (1974) with hamadryas baboon triads, a male–female pair and another male. In these experiments, it was the male possessor of the female who notified the other male. Abegglen argued that in the natural habitat, notifying should correspond to this sequence: a unit leader should notify other adult males, it was from these that they had mostly to fear an attack. He also suggested that it should be relatively new leaders (initial leaders), those in possession of one juvenile female only, that would be most competitive (yet most under threat) and would therefore be notifying more. During nine and a half months of observation, Abegglen recorded incidents of notifying, and analyzed frequencies of notifying and "receiving" between unit leaders (mature males with one or more adult females), initial leaders, and bachelors.

Table 3.1 gives the results of a systematic sampling of social contacts. It clearly shows that initial leaders are mainly the recipients of notifying, and notifying is done mostly by unit leaders and initial leaders. This supported Abegglen's prediction that harem leaders should primarily be the notifiers.

Initial leaders were most frequently both notifiers and recipients of notifying. Why was this so? Abegglen's explanation was as follows, initial leaders, possessing only one juvenile female, would be on the look-out for more females. They would constitute a threat and the unit leader would, therefore, "notify" to them. Initial leaders, however, were also under threat from each other and mutual notifying was frequent.

The meaning of notifying to its recipient was now shifted by Abegglen from something to do with changes of location and direction, to efforts at pacifying a rival male. Of course, it could have both meanings—as will be discussed in a moment. Abegglen had shown that the male who was likely to be challenged, did the notifying. He mentioned that initial leaders, after gaining adult females, were afterwards notified less because they were now less likely to be challengers. He also observed that a defeated unit leader stopped notifying after he lost his females.

Divergent Interests and Notifying

Abegglen had not rejected Kummer's earlier line-of-march explanation, but qualified it by the further explanation that before the march began a leader needed to notify possible rivals. During the march, units would become more dispersed and therefore vulnerable. Take-overs of females took place on the foraging march rather than at resting times. The leader's position became "more challengeable when his unit (was) on the move; then the leader's attention was diverted" (Abbeglen, 1984, p.154).

Abegglen added the following comments on four aspects of notifying:

1. Notifying was done only between adult males, that is, between individuals with low *a priori* compatibility.
2. As maturing males began to groom females and to search for females, notifying also tended to begin.
3. As argued earlier by Kummer (1975), notifying reduced hostility in the recipient. This is consistent with the use of the presenting gesture, a central feature of notifying, as a gesture leading to positive relationships.
4. Notifying was not at all evenly distributed between males but it was performed by possessors of females and addressed to rivals. Males as notifiers would be the motivated to prevent a fight, to be friendly. Notifiers tended to be males who might lose more in the event of a fight.

Abegglen ended with some speculations. Notifying occurred when males took initiatives in movement, direction, and, possibly, speed of travel. Both before and during the march, notifying could happen because of tension, either over females or about the direction of march.

The development of notifying explanations in this way provides a good example of how the close study of signals can increase our understanding of their meaning. The social context and the identities of both the sender and recipient of the message had to be taken into account. More recent research has concentrated on the content of communication between individuals. The work of explanation had narrowed its focus from the socio-ecological aspects of group structure to the socio-psychological level of joint actions in dyads.

Locus of Meaning

The locus of meaning had now been moved from group structure to social relationships, from one level of explanation to another (Hinde, 1997). Observational studies at Madrid zoo of hamadryas and yellow baboons (*Papio cynecephalus cynocephalus*) and their hybrids have since produced further information on the micro-signals of gestures, facial movements, touch, and vocal sounds from video analysis (Colmenares, 1990, 1991). Notifying encounters are far from stereotyped and are a striking example of how facial expressions, vocal sounds, touching, and changes of posture combine in the rapid signalling of baboons. Breaking the code, uncovering more of the meaning of baboon signals, has been one result of this detailed analysis.

There were many variations in what, at first sight, had seemed to be a fixed pattern: eye-contact, approach, turn, touch, withdraw. Sequences vary in the degree to which the signals from each partner match each other. If the initial eye contact is not accepted, the greeting approach does not happen. Where there is tension, touching is avoided and the greeting left incompleted. Notifying is now to be understood as a series of joint decisions, almost like questions, answers, refusals to answer. The messages, accepted or refused, carry consequences for the relationship. What is most interesting is the flexible control of various signals that simultaneously involve vocalizations, facial expressions, touch, and posture. Expressing this more technically, baboons use multimodal signals and appear to use a high degree of flexible control of complex sensorimotor skills.

Conclusions on Notifying

A feature of the studies of notifying is the researchers' explicit attention to the immediate social context of actions. In the studies by Kummer and Abegglen there is much information relating to the cognitive abilities of baboons in the context of the foraging march and the avoidance of conflicts. In the studies by Colmenares, attention is given to the use of signals influencing the social relationship.

Inferences about the meaning of notifying were elaborated as more information was collected; as researchers examined the occasions, circumstances, and participants in each notifying incident. Finally, the interpretation of notifying was based on the observation of signals having intended effects on relationships. The zoo study introduced the concept of negotiation, involving offers, attempts to persuade, and assessment of another individual's offers, the subject of the negotiation being the relationship itself. Are the signals, once made, also accepted by matching, answering signals? Is eye-contact held, is touching offered and accepted? How significant are signals like eye-narrowing, ear-flattening, and vocal sounds? Do they affect choices and decisions? Studies of baboon alliances have suggested a process of negotiating offers, acceptances, or refusals, and have provided new insights (Nöe, 1990).

In this search for the meanings of baboon greetings, two levels of explanation were used, one appropriate to group structure, the other to the relationships of dyads (Hinde, 1997). The units of analysis were, the structure of group (the troop) as well as the social relationships between males (the dyad). Initially, notifying was interpreted as functional for social organization at the structural level of the group. Later, explanations were offered at the level of dyads. Signals were now considered for their consequences affecting friendly contacts and avoiding competition. This approach encouraged a closer examination of the signals exchanged, as has been done in studies of baboon alliances (Nöe, van Schaik, & van Hooff, 1991).

The proximate explanation that signalling is analogous to a negotiation is a pathway to understanding more about the social intelligence of baboons. Meanwhile, at the functional level, the greetings are of interest if their result is the cohesion of the group and if ultimate reproductive advantages derive from friendships or alliances. Proximate information sustains theory at this level too. There is also a dialectic in the convergence of the two levels of explanation. On the foraging march, notifying can be understood as maintaining friendly relationships in dyads while serving to hold the travelling group together. During travel, the strains on the group are greatest and leaders are at their most vulnerable because their units of several females, infants, and juveniles are scattered. At this moment, the two explanatory elements of structure and relationship can be understood as complementary and interacting with each other.

CHAPTER FOUR

Facial Expressions

INTRODUCTION

The use and meaning of facial signals in nonhuman primates are under-researched topics. In humans, the importance of facial signals has been well emphasized as even greater than that of language in contexts where the message is about emotions or intentions. There is no equivalent work as yet on monkeys and apes, identifying the facial "action units" that Paul Ekman and Wallace Friesen (1978) and others have put under close scrutiny in research on human facial expressions, but there are some promising beginnings. Fine-grained analysis is rare but is starting to appear in the literature.

Up to the late 1970s the study of primate social communication was concerned largely with visual and gestural displays including facial signals, and in the 1980s and 1990s the emphasis of research shifted to vocal signals, because of the development of technology for the analysis of vocal sounds. A survey of articles published in the journal *Animal Behaviour* on the social communication of nonhuman primates between 1990 and 1995, found that 90% of them focused on vocalizations (Maestripieri, 1996a). Studies and reviews by van Hooff (1967) and Redican (1975) were historical landmarks in the study of facial expressions; much of the current work acknowledges their contribution to classifying facial signals across primate species.

Let us be clear on a question of terminology before we continue discussing the face as a source of communication. The viewpoint that facial expressions in humans involve merely the social transmission of emotions has been strongly contested (Chovil, 1991; Chovil & Fridlund, 1991). These authors propose the term "facial displays" as more accurate than "facial expressions" because these displays are better understood as communicative acts serving social motives. From a social psychological point of view, this preference is sensible, although we should also agree that emotions and sociality are often inseparable (Buck, 1991). Here, in this chapter, I use the terms signals and displays as well as expressions. Many of the researchers whose work I discuss refer to facial expressions without attaching to their descriptions any particular assumption of emotions. They have usually been more concerned with functional effects.

Facial movements, as with other channels of communication, carry important messages about current and subsequent actions and can control the actions of the receivers of the signal as well as being a principal channel for emotional displays, either honest or intending to deceive. It is important to avoid an overemphasis on emotions in the explanation of facial movements. There is the well known work by Ekman (1982) that has classified human facial expressions as reflecting basic emotions, but I also follow here a "behavioural-ecological" view, a term that Fridlund (1994) used for an approach emphasizing the communicative purposes of signals. This does not mean that I discount the pioneering analysis of facial expressions that Ekman and others such as Izard (1977) have contributed.

Fridlund (1994) argued that facial signals cannot only be derivatives of a few fundamental emotions. In humans, a cry-face might be put on because an individual is happy, angry, frightened, or relieved, but in all cases the aim is to receive attention. His view was that facial expressions are not usually indicators of emotion so much as instrumental devices to negotiate social encounters and facial displays should not, strictly speaking, be called "expressions" but "declarations" (1994, p.130).

Much of Fridlund's argument concerned the difficulty of attributing any one emotion to a facial expression and the further difficulty of "fixing" any one emotion to a particular facial pattern. There are yet graver difficulties in deciding on the criteria for determining which facial movements represent blends of emotions. Thus, he argued that facial displays can be satisfactorily interpreted in the context of ongoing conversations rather than by trying to categorize them by distinctive emotions. In humans, a relatively small proportion of facial signals are, in fact emotional displays (Fridlund, 1994; Russell, 1994). In humans, voluntary (controlled) expressions originate via the motor cortex, whereas involuntary (spontaneous) movements originate principally via the subcortical region (Rinn, 1984). Facial signals, in humans, are clearly not limited to the spontaneous kind and in social situations they are posed, managed, and, occasionally, deceptive. We do not find versatility to such a degree in nonhuman primates but facial signals are still very much involved in their social stratagems. The facial expressions of nonhuman primates have been *terra incognita* until very recently but, as this chapter shows, our knowledge of them is being rapidly extended.

Among monkeys and apes, both visual and vocal signals are major channels of communication but they are of varying relative importance. Amongst ground-living primates, for instance, visual signals are probably the most important.

Estimates of the proportion of baboon signals that are not visual but vocal, range only between 15% and 25% and, for rhesus monkeys, estimates are even lower, at around 5%. In certain situations, for chimpanzees travelling through the forest and generally for the arboreal species, vocal calls are relatively more important. Alarm calls are vitally important in contexts of danger but calls informing others about food sources or simply for keeping in contact with others are also important. The demands of the forest environment on the New World monkeys of South America also operate on some of the Old World monkeys in Africa and Asia which use more vocal than visual signals. Arboreal species tend to be less visually expressive than macaques and baboons.

Many of the New World species have only a limited range of facial expressions. The titi monkey, for example, communicates very little by facial expressions. However, in other New World species such visual signals are certainly important at close range. Marmosets (*Callithrix* species) threaten each other by eyebrow raising and grimaces. Capuchin monkeys (*Cebus* species) have a repertoire of facial expressions similar to those of Old World monkeys, including grimaces with teeth-chattering, head-shaking, blinking, and scalp movements (Fedigan, 1982).

GROUND-LIVING SPECIES: FACIAL SIGNALS

The hairless faces of chimpanzees and bonobos are the most expressive of the nonhuman primates. It should be relatively easy to read their emotions or intentions but they can also be very effective deceivers. Baboons' facial expressions, because of their thick fur and deep-set eyes, are not so easy to discern, especially when there are rapid glances and eye movements. Video analysis is one way to solve this problem, although in the wild this may not be practical. Field researchers have usually relied on their proficiency as experienced observers in reporting and interpreting the signals. One probable reason why facial expressions have not received more attention is the sheer difficulty of observing them in the wild. In captivity, the extreme rapidity of many facial signals expressed between two or more individuals has made it essential for researchers to use video recording.

In any conversational exchange, several modes of signalling are used. A male baboon chomps (opening and closing his mouth), grunts, glares, flattens his ears, raises his eyelids and brows, makes sudden body movements and rushes. Female baboons make threats by teeth grinding, yawning (showing the canines), and repeated circling. When Darwin (1871) used photographs of an actor to illustrate human facial expressions, the actor was also shown using hand and body movements simultaneously. Reporting facial, vocal, or gestural signals separately loses much of the actual process of signalling and therefore of communication. Several social psychologists have held the view that, in human conversation, gestures and speech "are parts of the same psychological structure" (McNeill, 1985, p.350) and are integrally linked in semantic and pragmatic usage.

SIGNALS AND TRANSITIONS: BLENDING OR GRADING IN FACIAL EXPRESSIONS

Researchers have emphasized that primate facial expressions blend into each other and that there are always transitions. This has been the view of every researcher who has tried to categorize them. As Alison Jolly (1985, p.199) remarked, "primate facial expressions are appalling to classify." This is partly due to the compound nature of expressions (using different muscles and areas of the face). Transitions constitute a problem for any effort to classify but it should be no surprise that facial expressions combine and change, nor that the meanings of these changes should differ from one species to another. There are blends, mixed and fleeting expressions. In some human cultures, for example, raising the brow (surprise) with the rest of the face neutral means doubt or questioning; if it occurs with head movement sideways it is an exclamation; if the surprise brow and eyes are combined with a happy mouth, this can mean enthusiasm. These transitions and combinations are, incidentally, one reason why it is often difficult to be sure about the emotions of the signaller (Ortony & Turner, 1990).

Lists of facial expressions in apes were compiled 20 or more years ago and it may have seemed that there was little more to add. It is only in the very recent work that attention is being given to facial signals in the dynamic context of behaviour. If social actions are to be considered for what they reveal about intentions, facial expressions should now acquire a new significance in research.

CATALOGUING HUMAN AND NONHUMAN FACIAL EXPRESSIONS

The interpretation of human facial expressions and their blends involves careful observation of rapid facial signals giving information about intentions as well as emotions. Some signals "flash on the face for a matter of seconds or fractions of a second" (Ekman & Friesen, 1975, pp.10–11). For purposes of analysis, there are three facial areas: brow (forehead), eyes, and lower face. Emotions may be expressed using only one or two areas or else the whole face, and transitions can reflect mixed feelings. Complicating matters further, the timing of a signal can reflect a mood, emotion, or intention: for instance, how long did it take for a pleased look to appear and how long before it faded? (Ekman & Friesen, 1975). Fleeting expressions are important, as these authors show, but it is worth emphasizing here again that, for a social psychology of primates, we seek to understand facial signals in the context of gestures, body postures, and vocal sounds, and, above all, in the relationship between the sender and receiver of the signal.

Earlier studies of baboons were invaluable for their lists of the facial signals of savanna baboons (Hall & DeVore, 1965). Signals were divided into three classes: (1) aggression (attack – threat); (2) tension (escape – fear – uncertainty); and (3) friendly approaches. Aggression was expressed by teeth-grinding, yawning (the threat-gape), staring, eyebrow raising, ear-flattening, and jerking of the head. Tension was reflected in staring, fear-grins, looking away, and sideways glances. Friendly approaches were shown by lipsmacking and ear-flattening. Ear-flattening varied in meaning according to the context. Another signal, raising of the eyebrows, could be a threat between males or a friendly gesture from a female to a male. Using the same signal to convey a different meaning, depending on the recipient and the relationship, has often been mentioned in primate studies.

Nine major facial expressions have been identified in macaques that are compounds of eye, eyebrow, mouth, and lip movements, often made with vocal sounds (van Hooff, 1967). The mouth was predominant in these categories: the tense mouth; staring with open mouth and teeth covered; the bared-teeth face; protruded lips, pouting; relaxed open-mouth (play face). The position of a monkey's lips depends on how frightened he/she is: the more fearful, the more the corners of the mouth are drawn back. Among the earliest researchers was Ray Carpenter (1934) and, in his opinion, the face of a gorilla was very expressive in the region of the eyes, through the raising or lowering of the eyebrows, or through opening or partially closing the eyes. The lips and mouth were considered even more expressive of mood or emotion. Others, comparing chimpanzees, gorillas, and orangutans, have concluded that gorillas are the least expressive of the great apes (Maple & Hoff, 1982), although there have been few recent studies of their facial or gestural communication (Gomez, 1996).

Jane Goodall's (1986) book on the chimpanzees of the Gombe Stream National Park in Tanzania, well over 600 pages in length, had barely three pages on facial expressions, despite her comment that chimpanzees' elaborate facial muscles produce a variety of expressions. She selected for discussion the "dramatic signal value" of the jaw-closed grin (lips raised, teeth uncovered, jaws closed)—an expression often made silently by a startled, frightened chimpanzee. This expression was ideal for signalling danger in situations where it was unsafe to make sounds.

It is certainly useful to have a clear idea of the repertoire of facial displays before considering how they are used in sequences of behaviour. Knowing the repertoire is a preliminary to exploring how individuals communicate with each other, to studying their tactics and, possibly, their thinking. Undeniably, ethograms (lists) of the displays themselves have to be the starting point. It may also be helpful to consider human facial expressions in more detail, as an exemplary model for methods that might eventually be applied in studies of monkeys and apes.

Human Expressions

Six basic facial expressions were identified as expressing the major emotions of anger, disgust, fear, happiness, sadness, surprise. The expressions consisted of movements of the eyebrows, forehead, eyes, eyelids, mouth, and lips (Ekman, 1982, 1992, 1994; Ekman & Friesen, 1969, 1975, 1982a). These authors stress that an observer reading the movements has to take note of multiple combinations and changes, for example:

1. Is the gaze direct or deflected?
2. Are the eyes steady or moving?
3. Are the pupils of the eyes small or enlarged?
4. Are the eyelids wide-open or narrowed?
5. Are there wrinkles around the eyes?
6. Are the eyelids blinking or relaxed?
7. Are the eyebrows in a high or low position?
8. Are they drawn together or lowered at the corners?
9. Is the forehead smooth or wrinkled?
10. Is the mouth open or closed?
11. Are the mouth corners up or down?
12. Are the teeth covered or uncovered?
13. Are the lips pulled back or pushed forward?
14. Is the upper lip curled?
15. Is the nose wrinkled?

Table 4.1 gives a summary of the itemization of the major facial features and their changes.

TABLE 4.1
Human Facial Features and Signals

Facial Features	*Changes, Signals*
1. Eyes	a. direct gaze b. moving c. pupil size, large or small d. wrinkles around eyes
2. Eyelids	a. wide open b. lowered c. blinking
3. Eyebrows	a. raised/lowered b. drawn together c. lowered at the corners
4. Forehead	a. smooth/wrinkled
5. Mouth	a. closed/open b. corners up or down; drawn back c. teeth covered/uncovered
6. Lips	a. closed/open b. tightly closed, compressed c. pushed forward (protruding) d. upper lip curled
7. Nose	a. smooth, wrinkled (with raised cheeks).

For other primates, chimpanzees, bonobos, gorillas, and macaques, signals are similar in the areas of eyebrows, eyelids, and, especially, the mouth. Uniquely human signals are wrinkling the forehead or nose and showing the whites of the eyes. In humans, the eyes and the region around the eyes contribute more to facial displays than in other primates. The white region around the iris gives more emphasis in humans to eye-movement, direction of gaze, and eye-widening, for instance, in expressing fear or surprise.

Facial Control

In humans and other primates, both emotions and intentions may be concealed to mislead an observer. There are plenty of examples of this kind of deception in studies of chimpanzees, baboons, and other species. In general, monkeys and apes are skilful at suppressing facial signals. Chimpanzees, macaques, and baboons, males and females, remain, in many situations, apparently unaffected by the threatening gestures of others. An impassive face, with avoidance of eye-contact, is evidently a way of disarming a threatening opponent. Threats or challenges are simply not taken up, they are ignored. If the threats are vigorous, the threatened individual must exercise voluntary control not to react. Similarly, male baboons in courtship situations look away at tense moments. Looking away is a disarming signal and can be conciliatory.

Expressing emotions that are not really felt seems to be uniquely human—this specific kind of pretend signal has not been identified in monkeys or apes as yet. Studies from Darwin (1876) onwards have distinguished between the full smile—the Duchenne smile that includes wrinkling of the skin around the eyes—and simulated smiles in which muscles move around the mouth but not around the eye: to smile with the mouth but not the eyes. I mention the false smile as a uniquely human form of deception, involving appraisals of others' states of mind with the purpose of influencing them. The vivid report of a chimpanzee controlling his (fear) grin by turning away from his rival (de Waal, 1982) and, with his fingers, pulling his lips over his teeth, impressive deception though it may be, is a lower-level operation compared to the use of false smiles.

The meaning of human smiles has received much attention (Ekman, Davidson, & Friesen, 1990). The Duchenne (true) smile is, technically, composed of the movement of the zygotic major muscle pulling upwards the corners of the lips, together with gathering of skin around the eyes and raising of the cheeks by the orbicularis oculi muscle. It is possible that even for such a distinctive expression as a smile, the interpretation of the facial movements involved could become as complex and fine-grained as the analysis of vocal sounds by spectrograph. Altogether, in the Facial Action Coding System, there are no less than 44 movement units, their intensity, laterality, duration, and timing being among the items included for measurement (Ekman & Friesen, 1982a). There may even be as many as 16 different kinds of smile or more (Ekman, 1992; Ekman & Friesen,

1982b; Frank, Ekman, & Friesen, 1993), considering differences in the movement units, but without considering the changes in interpretation that other simultaneous signals and situational differences can produce.

Although the difference between Duchenne and non-Duchenne smiles was confirmed by Ekman (1992), it has been questioned in other studies. According to Schneider and Unzner (1992) the absence of a full Duchenne smile could be the result of lower intensity of feeling rather than of deception. Currently, the distinction between the Duchenne smile and false smiles is still a subject for debate (Fernandez Dols & Ruiz Belda, 1997). It may be difficult to distinguish between the two with certainty, in which case the intention of the smile should be interpreted in relation to situational cues. So, a smile would be interpreted as being sincere because of the nature of the social relationships in which it occurs (Fridlund, 1994). Contemporary discussions of facial expressions recognize that descriptive situational accounts are necessary (Izard, 1997; Provine, 1997) and that "facial displays occur in a flow of situated action and can only be understood as features of that flow" (Ginsburg, 1997, p.380).

The Grimace, Fear-Grin, or Smile

The silent bared-teeth grimace with corners of the lips drawn back, is general to monkeys and apes when expressing fear or intending to appease. Grimaces or grins are made routinely by lower-ranking adults or juveniles towards a dominant individual, male or female. When monkeys grimace at a dominant it resembles nothing so much as a human ingratiating smile. Researchers disagree about whether males or females grimace more; they also disagree over whether males or female threaten more, across different species. Grimacing during copulation has often been reported in monkeys and in female chimpanzees and bonobos. At first meetings after some period of separation, grimaces are often used in greeting.

The grimace has changed from an expression of fear to a signal of greeting in at least three different evolutionary lines: in monkeys, the mandrill (*Papio sphinx*) and Celebes macaque *Macaca (Cynopithecus) nigra*, in chimpanzees and bonobos, and in humans (Rowell, 1972). Expressions similar to human smiles have been reported in greetings between gibbons. Male baboons grimace in friendly approaches to females. In greetings between chimpanzee males both dominant and subordinate males grin. Although macaques and chimpanzees often grimace silently, this expression intensifies with high-pitched screams in situations of intense fear.

It should be noted in relation to the smile or grin that the social contexts and purposes of seemingly familiar facial displays among monkeys and apes can be quite different from those in human societies (Preuschoft & van Hooff, 1995). Functional adaptations during the course of evolution may have created different forms, uses, and contexts for any given display even if the displays in question may be homologous (ultimately deriving from shared ancestors). As in vocalizations, facial displays are graded signals and we would expect transitions and blends. This may lead to a certain degree of arbitrariness in classifying or identifying a particular signal. It is currently more of a problem in interpreting nonhuman facial signals than in research on humans where there are at least two well known methods for quantitative assessment of facial movements (Ekman & Friesen, 1978; Izard, 1977).

There are many differences between species in the silent bared teeth display and many variations, for example, by lifting only the upper lip, (in some macaque species) and the lipflip, turning the upper lip inside out, in geladas (*Theropithecus gelada*). At the same time, folding of the skin below the eyes, and upward retraction of the scalp has been reported for macaques. Gaze pattern and posture accompanying the grin can also vary between species. The discussion of these variations by Signe Preuschoft and Jan van Hooff (1995) is notable for their references to Ekman's Facial Action Coding System in suggesting categories for displays. These researchers seem to be on the brink of an exciting new phase in extending our knowledge of the facial expressions of monkeys and apes.

In rhesus macaques, the bared-teeth display is almost always made only from subordinate to dominant individuals as a signal of submission (de Waal & Luttrell, 1985). Emphasizing the proximate level of explanation, there are questions about the degrees of intentionality that the use of this signal could reflect. If it is first-order intentionality then the signaller wants to influence the actions of the other, to reduce the likelihood of aggression. If it is second-order intentionality these subordinate monkeys are sending a message to dominants that they accept their lower status (Maestripieri, 1996b); perhaps we could even say, with intention to modify their attitudes. From a functional point of view, the bared-teeth display is an expressive appeasement signal, it effectively reduces the risk of attack and there is little more to add. But interesting questions about intentionality remain, at the proximate level of social tactics. How much intentionality should be inferred? It is advisable, on the existing evidence, to assume only that teeth-baring reflects a combination of fear and first-order intentionality (intentions to modify actions).

The open-mouthed bared-teeth signal also occurs in Tonkean and lion-tailed macaques (*Macaca tonkeana*, *Macaca silenus*) with staccato breathing vocalizations (Abegg, Thierry, & Kaumanns, 1996; Preuschoft & van Hooff, 1995). In Tonkean macaques the grin is not restricted to appeasement/submission situations but is used in an entire range of friendly contexts, in greetings, before grooming, and in play. In Barbary macaques, the silent bared-teeth display was never reciprocal (Preuschoft, 1992). After receiving the signal an aggressor would resume friendly interactions with the signaller; it was clearly disarming. The display acted to avoid conflicts, it also contributed to reconciliations, as a conciliatory grin.

A welcome discussion of grins, touching on questions about internal feelings and social relationships, has been provided by an observational study of eight

Barbary macaques in open-range captivity (Preuschoft, 1995). This research points to two major aspects of interpreting facial signals: the facial movements (their morphology), and their behavioural and inferred motivational context. The latter involves considering not only the presumed intentions or feelings of the sender of the signal but also the effects on the receiver (see Hinde, 1985), a social psychological emphasis.

It is most useful to consider the use of the silent bared-teeth expression, in relation to what happened just before and after, in social interactions. A typical sequence would be as follows: a dominant monkey, possibly threatening, approaches or looks at another, the second monkey grins and the sequence ends (simple appeasement) or else friendly interactions follow such as body contact or grooming, in which case the end-result is affiliation. A variation in the bared-teeth displays of Barbary macaques differs from the grin or grimace in that the jaws are wide open and the sender looks alternately at two other monkeys, shifting the line of regard from one to the other (Preuschoft, 1995). This tactic has been referred to elsewhere as enlisting behaviour (see later in this chapter) in forming alliances.

Grins often express submission, yet in Tonkean macaques the grin blends with a bared-teeth variation of the relaxed open-mouth face and the latter is used for affiliation or play, independent of rank relationships. In fact, there are no signals specifically reflecting dominance–submission in this species. During a study of Tonkean macaques in captivity over a period of seven years it became clear that the silent bared-teeth display, with mouth open in varying degrees, or closed, can be used during play but also to initiate interactions, to greet, to invite others to follow (Thierry, Demaria, Preuschoft, & Desportes, 1989). This research has usefully extended our knowledge.

Laughter and the Play Face

Unlike the grin, the open-mouth expression, in most species, acts as a metacommunicative signal, indicating that a play session is about to begin. In the play face, the mouth is wide open and mouth corners drawn back slightly. Gorillas, chimpanzees, orangutans, and gibbons, as well as stump-tailed macaques (*Macaca arctoides*), patas (*Erythrocebus patas*), vervet (*Cercopithecus aethiops*), and rhesus monkeys all show the play face. Although described most often in the play sessions of juveniles and youngsters, it is also an adult expression. The play face is accompanied by a jaunty approach when individuals try to enlist others to play. This display in Barbary macaques is almost always made by juveniles in playful contacts involving touching with the hand, gnaw-wrestling, and with lip-smacking sounds (Preuschoft, 1992). Usually, the display by one monkey is instantly reciprocated by another. As in gorillas, chimpanzees, and bonobos, Barbary macaques also make voiced breathing sounds, rhythmic panting noises similar to a soft laugh, during play sessions.

The relaxed open-mouth face has sometimes been referred to as the *covered-teeth play face* and the sounds similar to human laughter have often been mentioned (van Hooff, 1967, 1972). Chimpanzees and gorillas do not show their teeth in their relaxed open-mouth expression, a major difference from human laughter, but bonobos *do*, as well as making a panting laugh (de Waal, 1988). In this, bonobos are certainly closer to humans, although we should not seize too quickly on unique similarities: the showing of teeth in the open-mouth play face has been observed in the New World woolly monkey (*Lagothrix* species).

In mentioning the vocal sounds accompanying the play face, we move into the area of "vocal–facial" signals that van Hooff (1972) had earlier identified. To some extent there will always be an arbitrariness in discussing facial signals or vocal signals as separate channels, for the obvious reason that those signals are hardly ever used separately in social communication. We can continue now to discuss the topic of laughter.

Laughter in young chimpanzees occurs in the context of rough-and tumble play and tickling also produces laughter. As Provine (1997, p.172) has noted, chimpanzees laugh "almost exclusively during physical contact . . . during chasing games (the individual being chased laughs the most), wrestling or tickling." Tickling games, both for chimpanzees and humans, may help to form social bonds: "we only tickle and are tickled by those with whom we have close social relationships such as friends, relatives and lovers" (ibid., p.172). However, it should be noted that the vocalizations accompanying the play face, the rhythmic panting noises, are more closely tied to the respiratory cycle than is the case for humans (Provine & Fischer, 1989; Provine & Young, 1991). Chimpanzees only produce a single pant-like sound in exhalation or inhalation, whereas humans produce a number of notes in exhalation, with similar durations, at regular intervals (i.e. "ha-ha-ha"). Even so, human laughter is a stereotyped sound that contrasts sharply with the flexibility of speech. In fact, it is "a vocal anachronism, a kind of behavioural fossil, that coexists with modern speech" (Provine, 1996, p.193). In humans, laughter is a simpler form of vocal expression existing alongside the more complex forms.

Threatening

Facial expressions conveying threats are frequently mentioned in primate studies. Threat displays, with an open-mouth face and head-bobbing, are characteristic of rhesus and other monkey species. In baboons, the open-mouth expression is combined with raised eyebrows showing a white band above the eyelids. Flattened ears are also a threat or fear signal in monkeys, depending on the context. A direct stare is a threat signal in many primates, including humans. Reacting to this by turning the gaze away may be a similar stratagem to the social cut-off in humans (although similarity is not sufficient for assuming continuity). Glaring by chimpanzees is given emphasis with compressed lips; the bulging-lips face was described by van Hooff (1972) early on. A chimpanzee's gaze, without other signals, is not threatening.

Eyebrows

In baboons and Old World monkeys such as mangabeys (*Cercocebus* species) and guenons (*Cercopithecus* species), eyebrows are important in signalling. Exposure of the light or coloured skin of the upper lids (light blue in the case of pigtailed macaques, *Macaca nemestrina*) is made as a threat. Lowering of the eyebrows, as in the human frown, has been reported in chimpanzees and bonobos. When monkeys make threats, the eyebrows usually move upward. The message sent by this signal depends on the relationship with the receiver. Raising the eyebrows, for example, is a threatening signal between males but is an invitation to friendly contact from a female to a male. In olive baboons (*Papio cynocephalus anubis*), females invite males to approach by showing the white band over the eyes and at the same time lip-smacking. Barbara Smuts (1985) called it the *come-hither* look.

The Yawn

A yawn exposing the large (in males, very large) canine teeth is a threat display, or else expresses social tensions in macaques and other Old World species including chimpanzees and gibbons. Gorillas yawn when human observers are nearby, no doubt for the same reason. A yawn is much less of a threat than an open-mouth stare. In studies of baboons, *directed yawns* towards individuals as a threat signal were distinguished from *nondirected yawns* expressing stress (Hall & DeVore, 1965). Yawns in humans have been studied more than in nonhuman primates but they are morphologically similar to yawns in nonhuman primates only to the extent that the mouth is opened wide. In baboons, for example, not only is the mouth opened but the head is thrust back and the teeth exposed in a sudden upward movement of the upper lip. In humans, yawning occurs because of boredom, fatigue, or impatience. Yawns have been described as displacement acts and vacuum activity, but yawning is not well understood despite being a common form of behaviour (Provine, 1986, 1996). Yawns in humans do send a definite social message, which is probably why in some cultures individuals try to conceal a yawn, but this aspect of yawning requires further study (Fridlund, 1994). Certainly, for primates in general there is need for more extensive attention to yawning as a facial signal.

Fight or Flight

Mixed motives seem likely to be present in threat situations—whether to attack or to run away. There is a continuum of threat and submissive gestures. The head-bob threat is a clear example of the delicate balance between attack and flight (Rowell, 1972). The head makes two movements, one down and forward, the other up and back. The down movement is the beginning of a move forward; the up and back movement is the beginning of moving away. A monkey with less fear puts more effort into the down movement, a more fearful one puts more into the up-and-back movement. There are facial expressions to accompany these contrary movements. The down movement is made with the mouth slightly open, corners forward (an 0 mouth). As the monkey becomes more fearful and the up-and-back movement is more persistent, the mouth closes and the corners begin to be pulled back. Should flight not be possible, the bared-teeth grin will appear.

Such a sequence of mixed threats and approach signals has been reported between two female Java or long-tailed macaques (Zeller, 1991). Both showed interest in an infant and came closer with displays of nervousness (self-scratching). Figure 4.1 shows the females' actions, just after this nervous approach, in the *social contingency* format (see p.56). Female *A* roughly grabbing the infant, receives a threat from *B*. *A* grimaces (appeasement), jumps away, and gradually the signals become more positive. *A* gives (bared-teeth) grimaces with teeth-chattering, inviting an approach. *B*, after making a series of mild threats (head bobs, stares, body lunges) looks away, then accepts the teeth-chattering invitation, presenting herself for grooming. Presenting is a conclusive and reassuring signal. The sequence ends in a grooming partnership, a *mutual contingency*, positive for both females. Female *A* had used several mixed facial signals expressing threat and fear-appeasement (eyebrows raised and grimacing) followed by mixed fear-appeasement and invitations to approach (grimace and teeth-chattering). Female *B* after mild threats used the look-away as a transition to approaching. Facial signals are shown to be highly significant in this sequence.

Pouting, Lip-smacking, and Teeth-chattering

In some nonhuman primates, pouting includes raised eyebrows and a slightly open mouth with lips protruding, and it may be accompanied by the coo vocalization (van Hooff, 1967). Researchers refer to chimpanzees' use of this expression when begging for food and hooting at the same time; they mention pouting, in monkeys, by infants separated from the mother.

The lip-smack, with its facial and sound components, can take several forms. Among vervet monkeys it has a wet sound produced by bringing the lips together repeatedly. In other forms of the lip-smack, there may be protrusion of the tongue, with similar sounds. In some cases, the teeth may make contact with a clicking sound. All these versions are friendly or appeasing signals in mildly tense situations. Some variations are reported for baboons, notably in descriptions of jaw-clapping. Here, lip-smacking with protruded lips occurs as the jaws move up and down. In macaques, lip-smacking often alternates with teeth-chattering, rapid opening and closing of the mouth. It is a positive signal used

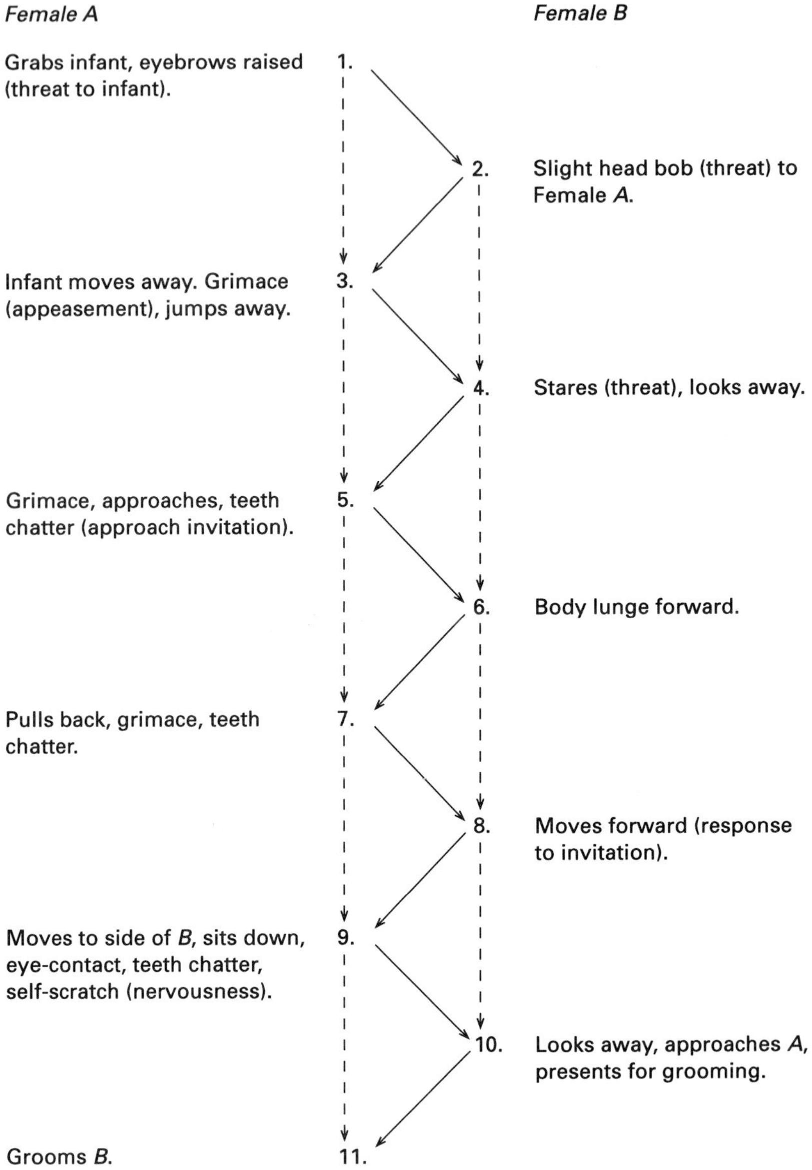

FIG. 4.1. Mutual contingency: Mixed signals leading to grooming (Adapted from Zeller, 1991).

in many relationships. For Barbary macaques, teeth-chattering shares the same proximate causes and effects as the human smile. Teeth-chattering in this species facilitates affiliation in the contexts of greeting, grooming, sexual behaviour, and interactions with infants (Preuschoft, 1992).

Teeth-chattering also occurs in the friendly approaches of Japanese and long-tailed macaques (Zeller, 1980). Teeth-chattering and lip-smacking are affiliative signals across macaque species: lip-smacking is general to all macaques. Teeth-chattering is derived from a blend of lip-smacking and silent bared-teeth grins (Preuschoft & van Hooff, 1995). In general, monkeys lip-smack in taking the initiative in grooming, in appeasing a dominant, in friendly approaches, and in courtship.

Head-shaking (head-flagging) has been observed to happen together with lip-smacking in baboons. This movement had been thought to express fear, averting the gaze away from a rival (Redican, 1975). More recent research on baboon coalitions has identified head-flagging as a signal to enlist an ally (Nöe, 1989; Packer, 1977). The enlister, by head-flagging, sends rapid glances to the potential ally and indicates the opponent. In similar situations, Barbary macaques make use of the unvocalized scream face, distinguished from the silent bared-teeth signal by widely open jaws and with "show-looking" towards two other individuals, quickly alternating the gaze, first towards one then towards the other.

The recent discussions of facial signals promise considerable advances in our understanding of social devices. Detailed information of these signals, in context, helps us to understand more about the nature and objectives of the stratagems in which they are used and about the degrees of cognitive skill used to produce them.

EYE CONTACT

Close Staring and Staring Down

A field study of over 200 close-staring episodes among six male gorillas shows that, in general, younger individuals tended to close-stare at older ones; it was occasionally an invitational signal (Yamagiwa, 1992). Other possible uses were in reducing tensions between adult males. There is a well confirmed contrast between these uses of gaze by gorillas and the gaze evasion by subordinates so often seen in monkeys. Interestingly, the gorillas' close stares were not accompanied by distinct facial expressions nor by physical contact. The stares were used relatively rarely but were definitely successful in encouraging friendly interactions.

There is also the case of *peering* in bonobos. Peering is defined as a prolonged gaze within 30cm. It usually leads to tolerance of a younger adult's actions by an older one. It is similar behaviour to the social staring of gorillas that we have just mentioned. Peering does not occur in tense situations except in encounters between groups. It is seen mostly during feeding and resting (Idani, 1995). Looking into the eyes of another at close range acts as a distinctive positive signal, in contrast to the *staring down* used as a supplanting tactic by dominant gorillas and baboons. Adult chimpanzees and gorillas are peered at by younger ones; in both chimpanzees and gorillas the *social stare* is made at closer proximities than when staring is used as a threat.

Close staring and staring down are accompanied in these social contexts by other facial, postural, gestural, locational, and vocal signals. As always, such signals are used in combination. Peering in bonobos does not involve the mutual eye-contact that often occurs in humans. There is an unanswered question about why the receiver of the signal does not peer back. Is it because, in this special case between older and younger bonobos, the older one merely tolerates the other at this point and the sequence of gestures has still to be amplified by touch or other actions of the younger individual? We would need to study the signal as part of a sequence of other signals.

So, to summarize: distance is an important marker between the *social stare* and the *stare down*. Eye contact itself is a centrally important element in social skills as indeed is the avoidance of eye contact—the decisive look-away that is so important in baboons during the initial approaches prior to forming a friendly relationship. In humans, gaze aversion has been shown to be a means to reduce physiological arousal (Gale et al., 1978). Eye contact has been interpreted as a way to express and assess intentions, both in humans and apes (Gomez, 1991). Eye contact, after all, is a major channel for communicating a great range of messages, positive or negative, which will determine the nature of a social relationship. Staring down in the context of threats is only one of the many uses of eye contact. Especially in apes, eye contact accompanies friendly approaches whereas often in monkeys the look-away achieves a similar purpose in initial friendly overtures.

For chimpanzees, eye contact also has an important use in reconciliations, in fact, mutual eye contact is nothing less than a prerequisite for a reconciliation. In contrast, rhesus monkeys, approaching each other after conflicts, look in all directions except the face (de Waal, 1989a). In apes, the use of eye contact is very similar to its social uses in humans. Apes use eye contact instrumentally to express intentions and to direct the attention of others to objects and events (Gomez, 1996). Close staring, into the eyes of a possessor of food, was a classic case in the early literature of this kind of usage among chimpanzees (Goodall, 1968).

Line of Regard

Using the eyes to follow the line of regard in another individual's orientation of gaze, tracking this line into space beyond, may be a precursory step towards developing a form of *theory of mind* (attributing desires, intentions, and, at its fullest development, beliefs to others: see Whiten, 1993). Baboons did this in one of the examples of deception in Chapter 2, when a juvenile distracted his pursuers by gazing into the distance. However, as Povinelli and Preuss (1995, p.423) make clear, although a chimpanzee can track the direction of gaze of a human trainer, looking at some external object and "young chimpanzees are sensitive to the eyes of others, they do not appear to understand them as portals through which the state of attention emanates." In an experiment in which chimpanzees were trained to use their natural begging gesture to request food from either of two trainers, one only of whom was blindfolded, they did so randomly. Nevertheless, the ability to follow another's gaze, which researchers refer to as joint visual attention or deictic gaze ("looking where someone else is looking": Butterworth, 1995, p.329) is considered an important advance in human infant development that "is thought to pave the way for deictic gestures such as manual pointing" (ibid., p.329).

RECOGNIZING FACES

The ability to identify individuals' faces has been demonstrated in experiments with chimpanzees (Boysen & Berntson, 1989) and rhesus monkeys (Keating & Keating, 1982). The eyes are shown to be a primary recognition cue in baboons (Kyes & Candland, 1987). Recording of eye movements in an experiment with four rhesus monkeys established that they fixated on the area of the eyes in human Identikit photographs. Changes in the Identikit arrangement of the eyes disrupted recognition, as did changes in eyebrows, to a lesser extent. Changes in the lower part of the face had no apparent effect (Keating & Keating, 1993). The invasive techniques used in this experiment, with a scleral coil inserted around the eye to measure its movements, do raise ethical questions about whether or not the results justify the effects on the monkeys. As other experiments illustrate, it is possible to use methods that avoid such dilemmas.

Research also indicates the importance, for facial recognition, of eyebrows and mouth as furnishing additional cues. For Japanese monkeys, important cues in recognizing facial expressions are "thrusting the mouth" and raising the eyebrows. In an experiment using bar-pressing with sweet potatoes as a reward, four monkeys had to choose between two photographs, one of which matched a sample photograph alongside; one monkey was able to obtain a high level of correctly matched responses (Kanazawa, 1996).

CONCLUSIONS

The ethogram or descriptive list approach has been important in the past in showing the range of primate facial signals. Advancing from that, more information is needed about their use in social interactions and only in very recent studies is such information becoming available. We especially need more information on the morphology and uses of facial movements and the transitions between facial expressions.

Many discussions of primate facial expressions have been concerned with their evolution or origins. Their use in social situations has been much less of a concern. Only a few steps have been taken towards studying their meanings in sequences and conversations of signals. So far, we have seen that the areas around the eyes and mouth are especially important and that facial signals are not used on their own, but combine with signals sent by sounds, movements of the limbs or body, and, possibly, by touch and odour. The importance of combinations cannot be overemphasized.

Facial signals play a part in the dynamics of relationships. Moods and intentions change quickly, from mild threat to outright attack, from wariness to friendliness. Signals are fluid and changeable. By trying to uncover the distinctive social uses of facial signals across primate species we become more aware of their communicative purposes. Our knowledge of human facial expressions is, at present, far ahead of equivalent knowledge in nonhuman primate studies. Empirical studies up to now have supported the case for the homology (common origins) of primate signals (Preuschoft & van Hooff, 1995). For nonhuman primates, I have considered mainly the smile (bared-teeth grin), laughter, the play face, facial threats, yawning, eye-contact, and the friendly signals of teeth-chattering and lip-smacking. These invariably occur with other signals, their meanings and purposes embedded in the immediate context of relationships.

Accounts so far of these signals seem like contributions to a progress report rather than leading to definite conclusions. In this chapter, my aim has been to assemble the work on facial displays, referred to by Fridlund (1994) as *terra incognita*. We are beginning to have a more comprehensive view of the flexible uses of facial signals in social tactics, enabling us to know more about how the signals act as instrumental devices. The value for researchers of compiling this information as a means to discovering more about animal cognition has been well stated by Marc Bekoff (1995). Discussing, for example, the inference that a monkey showed surprise, Bekoff (1995, p.134) asks the question "What would be convincing data?" and offers as his answer that "there is simply no substitute for detailed descriptions of subtle behaviour patterns that might indicate surprise—facial expressions, eye movements and body postures." In the following chapters, the available evidence on gestural and vocal communication in monkeys and apes will be examined for what it can show us of intentions and tactics in their social behaviour.

CHAPTER FIVE

Gestures, Postures, and Touch

Monkeys and apes communicate by vocal sounds, gestures, touch, and facial movements. Exchanges of signals between individuals seem similar to human conversations to the extent that a message is sent and the receiver sends a message in reply. Without a spoken language, the analogy to human conversations must seem restricted. Yet, in everyday conversations, messages, as spoken words, are involved with nonlinguistic signals: paralinguistic sounds such as murmurs, sighs, gasps, grunts, laughs, and the nonverbal signals of eye contact, facial expressions, gestures, and body posture (Argyle, 1988). In nonhuman primates these nonlinguistic signals are all present and often the vocal sounds take on referential meanings which, at their most interesting, signify social events and purposes. Such elements justify our exploration of the social uses of signalling. Although, in human conversations, the paralinguistic elements can be thought of as subsidiary to speech, they are often decisive in determining the meaning of an utterance to the receiver. In the following narrative: "Bill gave a grunt of assent and grinned at her and said he thought all was well and that results were good", Bill has sent messages expressing contentment even before he speaks. His later words only confirm the main statement that his other signals have already made.

The signals of nonhuman primates are astonishingly similar to this type of nonverbal communication. We will now begin to examine signals by gesture, posture, and touch.

A CHIMPANZEE CONVERSATION

When chimpanzees communicate, they certainly convey emotions, but there are also attempts to influence others' actions, to coerce or persuade. We can now examine an example of a sequence of chimpanzee signalling. Details of primate social interactions are scattered through the literature and occasionally short sequences are reported in full as in the descriptions of baboon (Smuts, 1985) and rhesus monkey social tactics (Judge & de Waal, 1994). One of the very few available *detailed* descriptions of *prolonged* social interactions between chimpanzees concerned an adult male ingratiating himself with another by facial expressions and gestures (Bauer, 1979). This happened during a temporary reunion of a group in the wild. On these occasions adult males often carried out menacing charges followed by grooming sessions which evidently restored friendly relationships. The majority of charging displays were made by the alpha male, Humphrey. Other chimpanzees reacted to his displays in a variety of ways. They either ignored the displays, pant-grunted, moved away, or crouched. One male, Figan, tended to delay his reply more than the rest. Humphrey was frequently observed continuing the display until Figan eventually gave submissive signals or left the scene.

The sequence given in detail in Fig. 5.1 happened at the end of one of Humphrey's displays. At this point, Humphrey continued to send threatening signals by his erect hair and stiff gait. The display itself was over and an approach by Figan now began. The sequence is shown in the Figure in the form of the *social contingency model* which represents the effects of each individual's actions towards the other by diagonal arrows and each individual's goal path by vertical arrows (Chadwick-Jones, 1991). This model shows each of the 24 steps in the exchange. The sequence lasted from the end of Humphrey's display, to the beginning of grooming with Figan and it shows a high degree of coordination of facial and gestural signals as a *mutual contingency* develops, the joint objective being a friendly relationship.

Many similar incidents have been described although not with quite this amount of detail (de Waal, 1982). A great deal is known about chimpanzees' social behaviour in general as a result of the work of Goodall, Mitani, Nishida, Reynolds, Marler, de Waal, Wrangham, Boehm, Arcadi and others. Even so, accounts cannot be closed on chimpanzees' social skills (any more than on human conversations) when there is still much to be known about the subtleties of their gestures, facial expressions and vocalizing.

Humphrey			Figan
First minute, Actions 1–9			
H stands, hair erect, shoulders hunched, looking away from Figan.	1.		
		2.	F looks toward H, hair flat, holds his right elbow with left hand, left arm across his body.
H stands, hair *mildly* erect, looks away from F.	3.		
		4.	F avoids looking at H.
H, hair erect, begins to walk up slope, glances towards F.	5.		
		6.	F remains sitting, holding right elbow.
H, hair erect, walks up slope, behind F.	7.		
		8.	F glances over his shoulder towards H and holds right elbow.
H sits	9.		
Second minute Actions 10–11		10.	F turns, extends hand, grunts while looking towards H.
H sits, hair erect, scratching, begins self-grooming.	11.		
Fourth minute Actions 12–14		12.	F remains sitting, holding one arm with the other.

FIG. 5.1. A conversation between two chimpanzees.

Humphrey	No.	Figan
H self-grooms	13.	
Sixth minute Action 15 onwards	14.	F begins self-grooming.
H remains sitting, pauses in self-grooming.	15.	
	16.	F stands and walks towards H grunting.
H looks away from F, hair erect.	17.	
	18.	F has pout face, stands and extends hand palm down to H.
H sits looking away from F, hair erect.	19.	
	20.	F continues to extend hand and touches H's face with hand palm down.
H remains sitting, looking away from F, with hair erect	21.	
	22.	F crouches and pant-grunts in front of H.
H goes to lean on his elbow, with hair erect	23.	
	24.	At this moment, F gives full open grin and begins grooming H.
	25.	F's brother, Flint, approaches and puts his face close to H's face, sits and watches F groom H.

FIG. 5.1. (*cont'd*)

GESTURES AND SYMBOLIC THINKING

The Iconic Hand Movement

If the vocal sounds of monkeys are the remote forerunners of language, as it developed in human-like ancestors, then some of these sounds would be symbolic of something in the environment. Even better, the signal would refer to something out of sight of the sender or the receiver, as in the case of chimpanzee food calls or vervet predator calls, both of which are to be discussed in Chapters 6 and 7. Gestures can also carry symbolic content. A study of bonobos at the Yerkes Regional Primate Centre in Georgia, had the aim of finding out what gestures reveal about cognition (Savage-Rumbaugh, Wilkerson, & Bakeman, 1977). Chimpanzee hand and arm gestures involved begging, taking, and embracing motions but this study of three captive bonobos suggested more. Analysis showed that 21 separate gestures could be identified in pre-copulatory interactions. There was, for example, a good deal of touching with a lesser number of iconic hand movements representing and illustrating positions the signaller wanted the other individual to take. Studies of the bonobo Kanzi have also shown that he used iconic gestures to his trainers—twisting motions for example, to show that he wanted a container to be opened (Savage-Rumbaugh, 1984). This suggested representational thinking and anticipatory mental images, possibly a pre-language development similar to the proto-imperative signals of young children, pointing to something they want.

The study by Savage-Rumbaugh and her colleagues had an important feature in that it was nonintrusive. Whatever signals were produced were done naturally and could not have been the product of prompting by researchers. It could not be criticized for experimenter influence or "Clever Hans" effects, as have some of the chimpanzee language-learning experiments (Wallman, 1992). In the famous case of the horse "Clever Hans", he appeared to be able to solve questions of simple arithmetic and algebra until it was discovered he was picking up subtle, accidental cues from his trainer (Hassett, 1984).

Pointing

Chimpanzees typically do not point, although they use glances or body orientation to direct another's attention and have been seen to point with the whole hand. Chimpanzees in captivity normally reach with the arm and hand towards particular objects, gestures that are similar to pointing, but do not point with the index finger (Povinelli & Davis, 1994). Pointing has not been seen in apes in their natural habitat, although chimpanzees, bonobos, gorillas, and orangutans have all been trained in captivity to use referential pointing, sometimes with the index finger (Call & Tomasello, 1994). Pointing with a single finger or the whole hand to indicate an item of food beyond their reach has been reported in observations of three captive chimpanzees (Leavens, Hopkins, & Bard, 1996).

Evidence on chimpanzee pointing is still in process of accumulating, that much is clear. The bonobo Kanzi, in several experimental studies of word-learning, used indexical pointing (Savage-Rumbaugh, 1984). Bonobos seem to be unique among apes in the gestural use of pointing with the index finger.

Referential communication, as in pointing, or indicating with a glance or movement of the head (gaze alternation) can be understood as intentional behaviour involving the use of a social agent to achieve an objective. Referential communication by gaze alternation, for example, is considered significant in developmental studies of human infants. Not only is this type of communication widely confirmed in apes (Call & Tomasello, 1994) but also brings to mind baboons' rapid gaze alternation in recruiting allies.

Begging Gestures

Bonobos use more hand gestures and more feet gestures than chimpanzees. Three kinds of begging gestures were reported from a study at San Diego Zoo (de Waal, 1988). One of these was begging with one leg and foot stretched out. The other two kinds, involving the hands, were hand-up begging with the palm facing upwards and hand-side begging with the palm facing sideways. Due to such variations, the gestural repertoire of bonobos seems more complex than that of chimpanzees, even though there are currently few accounts of bonobo social episodes.

These bonobo begging gestures were interpreted as requests for social favours, such as grooming. Very few of the begging gestures (13%) involved food, although this was the principal context of chimpanzee begging in Goodall's (1986) studies. Chimpanzee females usually begged from males, because males were more likely to be the possessors of meat and because females were rarely rewarded if they begged from other females. Sexually active females were the most successful in begging from males, plausibly suggesting a transaction.

Bonobos' and chimpanzees' begging gestures are often used after fights, in reconciliations. The equivalent after-fight signals in rhesus monkeys are embracing and lip-smacking. In chimpanzees' reconciliations, presenting and lip-smacking are used in addition to the begging gesture with outstretched hand, palm down, as in the Humphrey and Figan sequence. This palm-down gesture, placing the back of the hand near to the face of another is evidently an appeasement gesture. Begging gestures are accompanied by touching of the individual's hand but in appeasement it is the face that is touched (Maestripieri, 1996a).

In studies of pigtailed macaques, touching the face with the hand is labelled "touch-face" (ibid.). Contemporary research in general rarely mentions this type of signal. In the case of the touch-face, the hand is used together with, or immediately before, the facial signals of lip-smack or pucker (protruded lips) which in this species are friendly signals. Touch-face may form a composite message with other facial signals or is used to get attention. If touch-face is an attention-getting signal, the cognitive implications are interesting because it suggests the ability to take another individual's perspective. On the other hand, it could be that monkeys have learned that visual signals are effective only if the other individual is facing them and touch-face is an effective way of achieving this.

Other hand signals in bonobos and chimpanzees are sometimes mentioned. Bonobos' wrist-shaking and lateral flapping movements of the hands were interpreted as impatient, nervous invitations (de Waal, 1988); these gestures have also been noted in chimpanzees by different observers but are not widespread in either species. Bonobos use arm-waving as part of the male sexual invitation to females and this becomes a stretch over of the arm, palm downwards, if the partner comes close enough. Despite similarities, bonobos' gestures seem to be a more subtle mode of influencing others and their hand gestures are more like human (infant) pre-linguistic signals with the purpose of getting another individual to do something, as in the case of their use of the iconic hand movements that were described by Savage-Rumbaugh (1984).

So many of the examples I have just given support an interpretation that recognizes the intentional uses of gestures. Not only that, but they also give insights into the symbolic uses of gestures, an extemely important aspect of the social interactions of monkeys and apes.

PRESENTING: ITS MULTIPLE USES

The presenting gesture occurs very frequently in monkeys. A group of pigtailed macacaques were observed in captivity for 100 hours and, from the subsequent analysis, presenting was seen as the most frequent gesture (Maestripieri, 1996a).

Presenting transmits a variety of messages. It also takes on different meanings from one signal context to another. We turn now to observers' interpretations of presenting; there is a range which includes sexual and submissive meanings and extends to its invitational and reconciling purposes. The social uses of the presenting gesture were reviewed elsewhere (Chadwick-Jones, 1989); I now summarize some main features.

In studies of nonhuman primates, a female is said to present when she turns her hindquarters towards a male prior to copulation but presenting is also very frequent in nonsexual contexts, between males, between females, or a male presenting to a female or a female to a male. Similarly, mounting can have a nonsexual significance. Often, when one individual presents to another, the reply is brief mounting, or a glance or touch of the hand.

In a field study of geladas, young females presented to a male leader as if requesting tolerance. If a new leader emerged, presenting from others clearly signified submission (Mori, 1979a). The routine use of presenting by subordinates when passing in the vicinity of dominants has often been noted (Maestripieri,

1996b). *Intention presenting,* a signal made from a sitting position and involving a slight elevation of the hindquarters, was interpreted as a permission-seeking signal in stumptailed macaques (Bertrand, 1969). It was used by a lower-ranking individual approaching a higher-ranking one. The same action, if not preceded by presenting, would probably have been punished. These forms of presenting were accompanied by facial signals, grins or pout faces and teeth-chattering.

Presenting is commonly mentioned in the context of greetings between adult male baboons, although the use of the term *greeting* can be criticized because it tells little about it as a social event. In species in which dominance contests are important (such as rhesus monkeys, baboons, and chimpanzees), the effects of presenting are similar to the principle of clarification in human greetings, the settling of the issue of rank. Rank is clarified by style of address, first name, last name, title, the use of *tu* or *vous* forms (and their equivalents, in many languages) or by other grammatical forms. In languages such as Japanese and Korean many parts of speech are marked to express deference, expressing the status difference between speaker and hearer (Thomas, 1995). Also, in humans, bowing, kneeling, or even crouching are expressions of deference. In chimpanzees, bowing movements by lower-ranking males contrast with a dominant male's efforts to make himself bigger by standing up with hair erect (de Waal, 1987).

In nonhuman primates, presenting and mounting become trivialized in the sense that they are casually used in social interactions. The nonsexual use of sexual signals has been well documented. Many investigators have noted the casual brevity of asexual presenting and mounting. Similar observations have been made about the asexual erection of the penis, in greeting embraces between male chimpanzees and in some monkey species.

Presenting by dominants to lower-rank individuals is often mentioned and is worthy of closer examination. Reassurance was usually the explanation favoured (Bertrand, 1969; Mori, 1979b; Ransom, 1981). The presenter's actions avoid a challenge, transforming the rank-related relationship into a process of mutual accommodation (Reynolds, 1981). This argument is supported by dominants' use of other reversed signals. An example is teeth-chattering, usually a fear signal, by a male rhesus monkey approaching a female (Preuchoft & Preuchoft, 1994). This was also the interpretation of the bared-teeth display by chimpanzees in similar circumstances (van Hooff, 1972). In other contexts, dominants may be seeking an alliance.

Presenting as Social Approval

When presenting was first interpreted as signalling social approval it was a very novel suggestion indeed (Hinde & Stevenson-Hinde, 1976). In social psychology, approval is considered an important item exchanged in everyday conversations, useful in all kinds of social relationships (Hinde, 1997). The notion that presenting signifies approval and that it is exchanged for other favours confronts us with a question. Can the reciprocated favours that follow from presenting be clearly identified? In descriptions of presenting to dominants, the return favour is often tolerance, allowing the presenter to do something that would otherwise be punished (Mori, 1979a). Immunity from harassment may be the reward. Tolerance of the presenter in relation to space, food, mates, access to infants, locations, are all likely favours.

In the social context of presenting, the nature of the relationship between the presenter and the receiver necessarily affects our interpretation of the meaning of the signal. The nature of the relationship provides a key to whether we interpret this signal as permission-seeking or as a reassuring gesture. A variety of changes in meaning are also reflected in differences in the posture, such as legs flexed or straight; by combinations with other signals, eye-contact, vocal sounds, or teeth-chattering; by the position of the tail; by the relative locations of presenter and receiver; or by the rapidity of presenting.

Early on, Hall and DeVore (1965) mentioned several different uses of presenting by olive baboons, each one with postural variations, and among them a form of presenting between adult females when the receiver of the signal was a mother with infant. In this case, the hindquarters were lowered by the presenter while looking over her shoulder at the receiver and lip-smacking. Rowell (1966) referred to the use of "flank-presenting" by baboons as an invitation to groom, emphasizing the uniqueness of the posture in this context. A range of such variations compel us to interpret differences in intention and of meaning for the receiver. A detailed account of variations in presenting postures is found in Struhsaker's (1975) field study of colobus monkeys (*Colobus badius*). Nonsexual presenting consisted of two kinds (Type 1 and Type 2), identified by position of legs, head, and tail, duration, and location relative to the receiver. Type 1 was very brief (two seconds or less) from a low- to higher-ranking individual; presenters were usually females or juveniles, receivers were most often adult males. Type 2 involved adult males as presenters and receivers and was usually preceded by threats to the presenter, followed by grooming in half the cases observed.

Presenting by yellow baboons was classified into no fewer than six categories by Hausfater and Takacs (1987) who distinguished between submissive, sexual, mixed sexual-submissive, affiliative, abbreviated, and grooming postures. This study drew attention to the distinctive cues for each category such as raising or lowering of hindquarters, grimacing, pitch of vocalizations, tail twitching or tail "neutral", tentative glances, and degree of reciprocity in the signals. Similarly, for mounting, such variations have long been noticed. Hanby (1972) had mentioned six main postures in mounting by Japanese macaques. One interesting category was "intention" or "symbolic" mounting when a monkey, approaching another from behind, placed its hands on the other's neck.

COMMUNICATING BY ODOUR

Signals recorded by observers as tactile may involve communication by odour. In both New World and Old World species, a monkey new to a group is sniffed and touched, visual recognition being made more precise by smell (Gautier & Gautier, 1977). Touching often involves smelling. This is confirmed in baboons' use of the nose and mouth in greetings with genital–stomach nuzzling, rump nuzzling, back-fur nuzzling, and mouth–head kissing. *Chemical communication* is the term generally used both for smell and taste. The importance of odour has probably been under-emphasized in accounts of animal behaviour. But for Old World monkeys and apes, smell and taste, equally, seem less significant than other signals. This does not mean that information given by smell is not socially important, as when chimpanzees and baboons mutually sniff hands during greetings. Odour obviously gives important signals at periods of peak oestrus, although in chimpanzees and macaques, the visual signals are more important. New World monkeys, in contrast, have a keener sense of smell. Among the social uses of smell, by the latter, are the scent-marking of branches and tree trunks, yet the effects of this in social behaviour are little known (Zeller, 1987). One New World species, brown capuchin monkeys (*Cebus apella*), frequently wash their palms and soles in urine, which may provide signals to others about individual identity (Phillips et al., 1994).

COMMUNICATING BY TOUCH

In human communication, physical proximity between individuals is an infallible reflector of the degrees of closeness in a relationship and of its formality/informality. As most people have close relationships with relatively few others, it is not surprising that in many societies touching is a rare signal compared to eye-contact and facial signals. Social norms are restrictive about who can touch and who is touchable.

There is also the matter of which parts of the body can be touched in everyday social contacts. In Western society even touching of a hand by a hand or touching of the shoulders is permissible only between close family members and same-sex friends, usually females. There are norms fixing precisely when and in what context touching is appropriate. Touching between adult males is very rare, except for a few specific contexts such as hand-shaking or embracing in greetings. Cultural rules produce variations, of course, but this is not our topic here. Conversational touching is accepted between women and between women and men (Hargie, Saunders, & Dickson, 1994). Touching is especially important in adults' interactions with babies and children and there is firm evidence of the therapeutic advantages of touch.

In nonhuman primates, touching seems less rare or, at least, more visible. In many species where individuals spend every moment of the day within closely bonded and familial groups this might be expected (see Appendix 1). But in these species too, touching, for adults, is mostly between females or between males and females. In species where male dominance is often contested by other males, such as baboons, when touching does occur it carries a highly important, positive message. When, where, with which others, does touching happen? What are the purposes of touching between unrelated adults?

The kind of touching that has received most attention from researchers has been grooming, carried out with the hands searching over the head, shoulders, body, and legs, parting the hair and taking away any small particles. This hygienic activity has developed a social meaning, calming for the groomee and probably for the groomer, helping to maintain social bonds. Grooming takes up a large part of a monkey's day when not foraging. In captivity situations where there is no travelling, foraging, or patrolling to do, it takes up even more. Infants, higher-ranking males, and females are the most frequently groomed in any group of macaques. In multi-male, multi-female groups, female monkeys do most of the grooming but males groom females during periods of consortship and in monogamous species, males groom females more (Seyfarth, Cheney, & Hinde, 1978). Chimpanzee males often groom each other, this is a significant exception to the general rule that adult males do not groom mutually and seems to act in reducing the tensions of rivalries and alliances among chimpanzees (de Waal, 1992).

Touch most frequently involves the hands. Baboons, males and females, touch with their hands in greetings and as a reassurance. In baboons, the area that is touched can be any part of the body including side, flank, rear, and, between males, the genitals. Adult male baboons often form alliances but relationships are tense because they are rivals for females. They are careful to maintain a safe distance from other males, and touching marks a change in the relationship, as was illustrated in the case of notifying (in Chapter 3). Allowing another male into such close proximity and to touch or be touched, communicates willingness to accept the risk of vulnerability (Smuts & Watanabe, 1990; Vasey, 1995). Honest communication of this kind involves potential costs and the acceptance of risk.

Chimpanzees often proffer a hand and the mutual touching of hands is very similar to human actions. Patting, embracing, and kissing have been observed during reunions of separated travelling parties. Another use of touching has been reported in gorillas, in situations of tension, when females put a hand on a male's shoulder or back; this may be an attempt to negotiate male tolerance (Watts, 1995b).

One captivity study of stumptailed macaques refers to hip-touch, hip-clasp, and genital touching as responses to presenting and approaching (Maestripieri, 1996a). Mutual touching is an important reassuring signal, important too in the tentative initial attempts to form a relationship. Figure 5.2 shows an interaction between two female olive baboons. Daphne is approached by a high-ranking female Psyche, and they touch each other on rear, hip, and belly (Smuts, unpublished data). Touching here is being used instrumentally to achieve a social purpose.

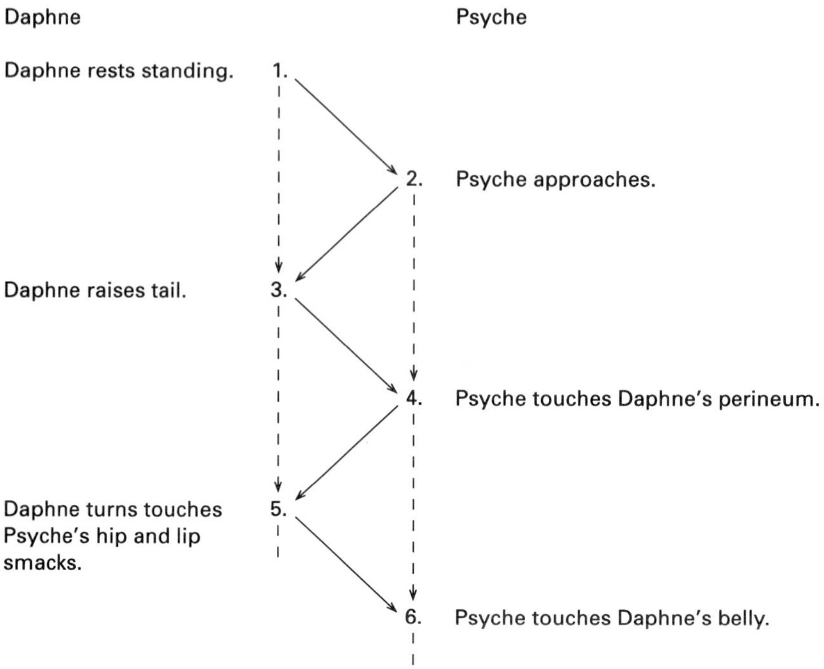

FIG. 5.2. A friendly interaction: Daphne and Psyche.

Touching as a Key Signal

One study in particular has given us valuable information on the use of touching. 637 greetings between adult male baboons were analyzed and were interpreted either as a negotiation prior to forming a coalition or as a test of relationships (and of relative dominance) without having to risk fighting. Although the range of conjecture about possible meanings of the signal is disconcertingly wide, at least it is clear that touching is very important for our understanding of the messages exchanged (Smuts & Watanabe, 1990).

Each greeting sequence was initiated by an individual catching the eye of the other in similar fashion to the greetings described by Colmenares (1990, 1991). This initial signal was followed by a distinctive approach with swinging gait, lip-smacking, grunts, and with the eyes-narrowed, ears-back face. Touching of genitals occurred at the moment when the two males were closest and after the approaching male had turned his rear towards the other. Less frequently, each male, in a mutual approach, touched the other.

A feature of the researchers' discussion of their results is their explicit treatment of the use of touch in the greeting. The risk-taking component in the greeting, allowing another male to approach so close, was an important factor because the initiating male made himself vulnerable to attack. Genital-touching from the male receiving the greeting could represent agreement, the culmination of a negotiated sequence; this is why I have called it a key signal. The action of approaching and being touched seems, in this context, to affirm commitment to the relationship.

In other situations, when two males combined to attack another they used brief versions of the same gestures to synchronize their joint actions. One pair of resident males, Alex and Boz, had a long-standing alliance, helping one another take oestrus females away from other males and protecting each other from rivals. This pair used the greeting sequence far more than any other, completing over 90% of them. Unfinished greetings evidently reflected tensions. This was the only pair with an entirely symmetrical pattern in hip grasping, genital touching, mounting, and in the initiating role (Smuts & Watanabe, 1990).

This study takes us towards a more adequate interpretation of signalling, mainly by the inclusion of signals by touch. In previous baboon studies these have merely been mentioned in passing. Yet, unless the touching signal was included in the sequence, the greeting failed. This was a necessary element in the exchange of signals, as important as eye-contact in other contexts. These baboon greetings can be compared to similar cooperative communications between male chimpanzees, where incompleted greetings might also be unsuccessful attempts at negotiation (de Waal, 1986b).

Negotiation and Touching

The sequence of signals leading to the formation of an alliance has frequently been interpreted as a negotiation process. Negotiation in this case refers to initiatives or offers, their acceptance or refusal. Certain specific signals are mentioned, such as head-flagging (rapid glancing and side-to-side head movements), thought to be important in enlisting an ally. In New World white-faced capuchin monkeys (*Cebus capucinus*), one monkey jerks its head towards a prospective coalition partner and then returns its gaze towards its opponent (Perry, 1996). In baboons it was a signal from the initiator of a coalition (Packer, 1977), although in other studies both partners made the signal simultaneously (Bercovitch, 1988; Nöe, 1989). Other baboon signals have been mentioned, in similar incidents, including presenting, mounting, touching, lip-smacking, and mutual patting. Current studies leave us with inconclusive suggestions about these messages and joint actions. The coalitions themselves may be the result of reciprocal altruism, or opportunism; no-one is fully sure. It will, in the future, be a matter of constructing further accounts of these episodes in baboon communication.

If an initial appeal fails, vocal sounds intensify (Nöe, 1989). Chimpanzees appeal to an ally by rapid glances, begging hand-gestures, agitated cries, even temper tantrums that seem to reflect intense negotiation (de Waal, 1992). No less than 15 distinctive actions and vocal sounds have been seen being made from participants in fights to potential allies (de Waal & van Hooff, 1981) but detailed attention to these signals, as they actually happen, is still awaited. Lists of signals are an excellent starting point; from there the search for meaning must take us into studying actions leading to definite consequences. The accounts of different researchers, focusing on similar episodes, are accumulating a pool of observations moving us towards consensus about their meanings. The question to be answered in situations of enlisting aid is: what kind of information is being sent or received? Our inferences about this can only come from detailed descriptions of how signals are used.

CHAPTER SIX

Vocal signals: Apes

INTRODUCTION

A social psychology of primates should emphasize the study of vocal sounds in contacts between individuals. Monkeys' and apes' voices not only express their emotions, they carry other meanings and purposes. Despite the fact that chimpanzees have been studied in the wild more than any other African primate, in the first 25 years of field research little quantitative research was conducted on their vocalizations (Arcadi, 1996). Chimpanzee vocalizing is receiving a great deal of attention now, especially with the aid of the sound spectrograph. Technical advances in acoustic analysis have assisted efforts to assess the uses of vocal calls in their social context (Zimmerman, Newman, & Jurgens, 1995). It has been believed for some years that vocal repertoires of monkeys and apes are much more extensive than had been thought (Feyereisen & de Lannoy, 1991) and new information on vocal signals, accumulating very recently, is giving ample evidence of this.

Vocal sounds send a message about the most likely next actions of the vocalizer and are a cue for the matching actions of the receiver; to approach, be friendly, or to appease, threaten, or flee. They influence or control the actions of others and express the nature of relationships between individuals, in both humans and nonhumans. If we remember that social psychology as a discipline, attempts to understand and explain how individuals act in the presence of others, then vocal signals and their influence on their hearers can be considered as of central importance for this approach to primate studies.

Early observers of nonhuman primates assumed that vocal calls were expressions of emotional states of the caller. This was the "groans of pain" view of vocal signals (Griffin, 1981). New evidence changed this view. It is now accepted that, at the very least, across different species, vocal signals carry messages about food sources, rival troops, and predators. It is also well accepted that the caller intends to have an effect on the hearer, in a variety of social relationships, as was discussed in Chapter 2.

GRADED SOUNDS

In chimpanzees, bonobos, gorillas, rhesus, Japanese, stumptailed, colobus, and talapoin monkeys (*Cercopithecus talapoin*), research has long established that the vocal repertoire extends over a graded acoustical continuum. As in human speech one sound blends into another and listeners encode the meaning of the graded sounds as entirely separate signals. Sounds that are acoustically similar may be perceived by other monkeys as distinct, as happens in human conversation. In earlier reseach, some complex graded sounds that probably sent separate messages to their monkey hearers could not be unambiguously classified as separate (Petersen, 1982). With more sophisticated analysis this situation is changing. A complicating factor is that the grading can occur along several dimensions such as pitch, tone, and duration, each varying independently. Spectrograph analysis and playback experiments with recorded sounds, are now helping to identify calls with slight changes of frequency and tone.

Some of the most interesting research has advanced our knowledge of the use of graded sounds. The effect of a graded repertoire of sounds is that it can convey subtle information. Much more remains to be discovered about the meaning of the fine gradings shown in the spectrograms of many studies, these gradations are potentially important because they involve the conversational, social uses of sounds. Researchers are able to distinguish some gradations by ear, others only by spectrograph analysis, but yet others, so far unidentified by the spectrogram, can be distinguished by the human listener (Boehm, 1989).

Chimpanzees' Graded Sounds

Vocal communication has been referred to as the least understood behaviour of chimpanzees, both in their long-distance and close-range sounds. Up to recently, there had been no systematic close-range study of the short-distance vocal communications of wild chimpanzees. Now that is beginning to change. One study, for example, describes how females at mating periods abbreviate their pant-grunts, a rather subtle change, as well as pant-grunting less frequently (Nishida, 1994, p.379): "It appears as if an adult female greets a male by saying 'Good morning!' when not in oestrus and by saying only 'Hi!' when in oestrus." This difference reflects the change that her condition has made to her relationship with a male. She can afford to be less deferential, she has something that the male wants and now the relationship can be negotiated differently.

Territorially defensive species need only one distinctive call to warn off rivals, whereas signals used in short-range contacts are graded and made with visual signals (Marler, 1976). Chimpanzees use both kinds of signal: loud calls to warn off intruding chimpanzees from other terrirories and yet a full range of graded vocal signals at closer ranges. The two most frequent chimpanzee vocalizations are the pant-hoot, keeping separate individuals in touch with each other, and the pant-grunt that sends a submissive message in dominance relationships (Mitani, Gros-Louis, & Macedonia, 1996). In the case of the pant-grunt, a sound with evident conciliatory effects, graded changes in this call reflect the social context, from tense situations to friendly greetings. If a subordinate individual is apprehensive, the pant-grunt grades to louder pant-barks that eventually can grade again into frenzied pant-screams.

There are many examples of grading in the pant-hoot, the most frequently heard chimpanzee distance call, voiced both in inhaling and exhaling. It identifies individuals to the human listener as well as to other chimpanzees. Research workers arriving in the field took only three weeks to distinguish between adult chimpanzees by their pant-hoots (Byrne, 1982). Differences in durations of pant-hooting were found to be related to age and dominance rank (Marler & Tenaza, 1977).

The scream of an excited or frightened chimpanzee might not seem a prime candidate for the encoding of subtle messages. Actually, there are differences in pitch, timbre, tone, duration, timing, which are apparent even to the human ear. Four types of scream have been identified: the victim scream, the tantrum scream, the SOS scream, and the copulation scream (Goodall, 1986).

Vocalizing with Others in Mind

Chimpanzee pant-hoots at food locations, which have a final roar-like ending, are not only associated with arriving at food trees but also occur more at empty food trees than at trees already occupied by chimpanzees. They have a social objective—to recruit others (Clark & Wrangham, 1993). Field studies have opened up new possibilities for interpreting these chimpanzee calls. Low-ranking chimpanzee males suppress vocalizing (when alone or with a female) to avoid feeding competition, whereas high-ranking males call loudly to attract mates or allies (Clark, 1993; Arcadi, 1996). Although female chimpanzees and low-ranking males tend not to make loud calls; high-ranking males are not so inhibited. Their calls are directed to allies or else to advertise group size (if there are loud answering choruses, the feeding party moves away). Low-ranking individuals, when in the company of high-ranking ones, do make loud calls, joining in the chorus. Probably, low-ranking chimpanzees on their own are discouraged from calling loudly because of fear of an encounter with a dominant. Whatever the intention of the caller, the dominant may understand it as a challenge.

Females and subadult males tend to respond to pant-hooting with calls other than pant-hoots; for example, with the whimper-hoots made by low-ranking individuals. Whimper-hoots are probably a tactic to avoid aggression from adult males. In these ways, researchers uncover the interesting influence of social relationships in vocalizing. Such interpretations of pant-hoots at food locations illustrate perfectly where we should seek the meaning of a vocal sound. It is, of course, in the relationship of the caller and receiver.

An important analysis has been made of male chimpanzees' pant hoots, in relation to allies, associates (other individuals who are usually in close proximity), and grooming partners. Pant-hoots, although they may have the effect of informing others about food locations, have further effects in forming and maintaining groups. In this study too, high rank encouraged pant-hooting and individuals hooted more often when associates and allies were close by but out of sight (Mitani & Brandt, 1994).

Thus, the loud pant-hoots send messages with a primarily social meaning. The call is referential, not just to signal the discovery of a fruiting tree. The call (and suppression of the call) relates to two kinds of social relationship. First, which chimpanzees not present might be hearers of the call? Second, which chimpanzees are currently present with the caller at the feeding site? Depending on either of these, the caller's signal may be made with greater or lesser vigour or possibly not made at all. The spectrographic analyses have shown differences associated with social messages and with relationships with other chimpanzees. The number of recent discoveries about chimpanzee vocalizing and the leads offered to further research are impressive.

Rough-grunting at a Food Source

In addition to the pant-hoot, another, less loud, vocal sound occurs in the context of food; this is known as the rough-grunt. The possible symbolic use of rough-grunts was discussed by Peter Marler (1977a) who discounted the alternative explanation that chimpanzees were signalling only their excitement at finding food. One investigator has found their meaning puzzling, considering it unlikely that these calls advertise the finding of food because they are too soft to attract chimpanzees from a distance (Nishida, 1994). It can also happen that chimpanzees make food-grunts long before they arrive at the food. The sounds might coordinate the feeding of several chimpanzees or they might still be expressing excitement. It has been speculated that food calls might refer to different kinds and quantities of foods (Goodall, 1986). Such questions about food signals are attracting current research.

Experiments with captive chimpanzees have also demonstrated that calls vary in proportion to the amount of food discovered. This research suggests further differences between the two distinct calls given in the context of food: rough-grunts and pant-hoots. Four captive chimpanzees were given different quantities of watermelon. The experiment showed that whereas rough-grunts were invariably uttered on food presentation, pant-hoots were only given in trials with the most number or largest melon pieces (Hauser, Teixidor, Field, & Flaherty, 1993).

VOCAL–FACIAL SIGNALS

As mentioned earlier, except for long-distance calls, monkeys and apes use sounds in combination with visual signals. This complicates the task of trying to understand purposive social behaviour, especially because it is rare to find accounts of the full range of signals as they occur together. Jan van Hooff's (1973) captivity study of chimpanzee body movements, facial expressions, and sounds, gave some indication of simultaneous signalling. For example, squat-bobbing (squatting and bobbing up and down), hoots and rapid "oh oh" rough grunts occurring at a rate of 3–4 per second, all happened together. Van Hooff's use of the term "vocal–facial" gave an emphasis to the simultaneous occurrence of signals, for instance, in the "stretched-pout whimper" and "bared-teeth scream". The bared-teeth grimace, silently, could be interpreted as a friendly signal, but with the scream or yelp, it was submissive. This early report gave some preliminary clues to the complexity of chimpanzee conversational messages. The use of vocal and visual signals, one modifying the other, has obvious similarities to multichannel communication in human conversations, especially for conveying subtle changes in a message.

DISTANCE SIGNALS

Loud Calls

Several sources refer to lower rates of calling (in comparison to males) by female chimpanzees and gorillas. Partly for this reason, females' calls have received less attention from researchers. Nevertheless, in groups with a variable fission–fusion structure (see Appendix 1), chimpanzees, bonobos, and the New World spider monkeys (*Brachyteles* and *Ateles* species), loud calls are given by both males and females (Hohmann & Fruth, 1995). Due to the large amount of fission and fusion that routinely takes place within a chimpanzee community, loud calls help individuals to keep in contact or find allies. Chimpanzees and bonobos are unique among nonhuman primates in making calls and choruses for friendly cooperation within a group, rather than just warning off rivals. A similar development may have taken place in human evolution (Boehm, 1989).

Loud calls, generally between individuals at a distance, are relatively less interesting because there is no joint use of visual signals and other subsequent social interactions tend not to be reported. There are differences in the uses of loud calls between, on the one hand, chimpanzees and bonobos and, on the other, orangutans and gorillas. In the former, loud calls are used by mature and immature individuals of both sexes to communicate with members of the same community. In the latter, the calls are restricted to adult males with spacing as a major effect, warning off rivals.

Gibbons and Distance Calls

Under the forest canopy visual signals do not reach very far. Some of the distance calls in this environment, are spectacularly loud and melodious. Such calls are often distinctive in their sound composition, without the grading of close contact sounds. Most species of gibbons, males and females, have daily morning duets, each with a different song. Playback experiments using recorded songs of male Bornean gibbons (*Hylobates muelleri*) have shown that the songs maintain territorial spacing (Mitani, 1992). If the playback speaker was located towards the centre of a listening male's territory, he immediately moved in the direction of the call. When calls were played from a neighbour's territory, males did not move. In these songs too there was a large range of individual repertoires.

Spectrographic analysis revealed a systematic tendency behind the order of notes that was reminiscent of human language and were an important feature in communication. While male gibbons responded by giving "squeak" calls (as a response to an intruder after hearing a normal song) they did not do so to rearranged playback versions (Mitani & Marler, 1989). This suggested that meanings depended on the sequence of notes in the song in a similar fashion as the meanings of words are determined by the order of phonetic elements (Mitani, 1992).

CHIMPANZEES' VOCABULARY

As more studies have been carried out, the preliminary list of chimpanzee calls (without taking grading into account) has become longer and more precise. Jane Goodall (1968) listed 24 calls, but later extended this to 34. Peter Marler (1976), in contrast, reduced her original list to 13, but the main elements remained the same and corresponded well to the earlier descriptions of gruff barks, pant barks, "waa" calls, chorus barking, and soft moans (Reynolds & Reynolds, 1965). The main signals were:

1. The hoo call at seeing strange objects;
2. The "eagle wraa", a harsh cry of hostility;
3. "Waa" barks when a chimpanzee was frightened;
4. Excited hoots;
5. Screams, squeals;
6. A variety of panting sounds during grooming, copulation, and in "laughing" during play.

There were also apprehensive pant-hoots or pant-grunts. Early summaries tend to emphasize emotional expression, whereas more recent research examines the evidence for intentions. The distinctive vocal features used by chimpanzees involve pitch, volume, duration, intonation, and repetition. There are subtle changes in single calls such as the pant-hoot and as well as the gradations from one call to another. There are dozens of acoustically different pant-hoots and grunts in a series that seem to be used as questions and answers.

VOCALIZING: GORILLAS' AND CHIMPANZEES' CONVERSATIONAL SIGNALS

Early studies estimated the frequency of chimpanzee vocal sounds as between 10 and 100 calls per hour in the feeding area at Gombe (Tanzania); a lower rate was likely among free-ranging individuals (Marler & Tenaza, 1977). More recently, it was estimated that individual adult chimpanzees vocalized only two to four times per hour in the unprovisioned locations of the Kibale (Uganda) forest, but more frequently in reunions with other chimpanzees (Clark, 1993).

Comparing rates of vocalizing for individual chimpanzees and gorillas, frequencies have been estimated for both as more than 10 times per hour (Marler, 1976). One long-term study gives the figure, for gorillas, of eight times per hour (Harcourt, Stewart, & Hauser, 1993). A second estimate for male gorillas at the same location (Karisoke, Ruanda) is in a similar range; two silverback males vocalized at around 15 calls per hour over 25 one-hour observation periods (Mitani, 1996). As noted, chimpanzee and gorilla males vocalize much more than females and this has to be a reason for the focus on male signals.

Research has revealed some interesting conversational possibilities in the sense that the sounds appear to be similar to asking a question and gaining an answer. Gorilla grunts (termed "double grunts" because of the way the sound is produced) can be distinguished acoustically from answering calls, a significant discovery, although it is presently unclear what message is conveyed. The first grunt may be a call settling the matter of rank if the answering call is in a submissive mode (to a dominant individual), or the calls may maintain spacing between individuals; there is no sure interpretation at the moment (Seyfarth, Cheney, Harcourt, & Stewart, 1994).

The most detailed analysis to date of apes' close-contact vocalizing comes from a study of two groups of gorillas as they fed, travelled, or rested in the wild (Harcourt et al., 1993). The study analyzed the acoustics and contexts of staccato *syllabled* grunts and *nonsyllabled* longer calls. These calls appear to operate to maintain contact or coordinate activities and to act as mild threats (syllabled calls) or appeasement signals (nonsyllabled calls). For the majority of the gorillas observed, about 60% of calls were part of exchanges, in a variety of relationships. Mothers responded to infants, adult males made calls in response to others' movements or when approached by others, and just before or after grooming. A nonsyllabled call (grumbling) seemed linked to submission and was most evident when a silverback male displayed.

The meanings of gorilla grunts, sounding the same to the human listener, needed often to be interpreted in their immediate context of other (social) indicators. There were three major contexts for these close-contact signals. First, any imminent change in spatial position, as at the end of a resting period (especially for syllabled calls). Second, when an individual was unusually near another (gorillas typically feed at a distance of more than five metres apart). Third, vocalizing in a chorus was not uncommon on finding or feeding on a large amount of food. So, close-contact vocalizing usually involved either potential separation or potential conflict (Harcourt et al., 1993). It is interesting that gorillas have no vocal signal that is the equivalent of the chimpanzee pant-grunt, used exclusively to indicate submission. Their subtle, close-contact sounds seemed sufficient for communicating in the small, static gorilla groups where all individuals are familiar to each other (see Appendix 1). In these circumstances, there is little competition over food sources and relaxed negotiation is the style of competition.

A further study of gorillas' close contact (syllabled) grunts, addressed the hypothesis that increased vocalizing at the end of rest periods was intended to assess readiness to depart (Stewart & Harcourt, 1994). An analysis was made of calls in 15 resting periods of two small groups of 12–14 members each. Gorillas synchronize their activities as a group. Most rest periods were between one and one and a half hours long. The study showed that vocal signals increased during the last quarter of the periods. Most grunts were part of an exchange. Many grunts drew a response from another individual, allowing the inference that the signal led to a consensus on the moment to leave.

The ongoing research on gorillas in the wild has thrown increasing light on true conversational aspects of vocalizing, where signals carry significant messages between individuals. Gorillas have a repertoire of at least eight distinct close-contact calls and there seems to be scope for discoveries about their social usage.

BONOBOS

The Vocabulary and its Intentional Uses

Research has also been carried out extensively with human-reared bonobos. The intentional uses of vocal sounds were brought into full focus by intensive studies of the bonobo male Kanzi, then an 18-month old. Kanzi used gestures and sounds to get his teacher to carry him from one area to another; to be lifted to a high place, to get the teacher to approach him, to help him in opening a bottle; to persuade the teacher to get him an object, and to appeal to one teacher to make the other teacher give him the object (Savage-Rumbaugh, 1984). These were all examples of vocalizing to control the physical or social environment.

Several studies of captive bonobos are interesting because of what they reveal of the social context of vocal signals. About 300 hours observation of 10 adult, adolescent, and juvenile bonobos was reported from the San Diego Zoo (de Waal, 1988). This has been the only bonobo study up to now to utilize spectrograph analysis. The aim was to compile a basic ethogram to serve as a guide for studies in the natural habitat. This preliminary analysis of the vocal repertoire yielded 12 categories. Worthy of mentioning is *contest-hooting*, a continuous high-pitched sound made in challenges between males. Hooting, mixed with yelps, grunts, or screaming formed a dialogue that apparently told each of the two contenders something of the intentions of the other. Dialogues were accompanied by invitational hand gestures, giving further cues to their interpretation.

In contest-hooting, the exchange was a dialogue rather than a "duel", in the sense that a dialogue involved turn-taking by both partners whereas a vocal duel meant blocking the other's signals and not conceding the turn (Todt, 1987). In contest-hooting, there was turn-taking. There were even examples of vocal matching, defined as mutual adjustments of sound structure. The essence of this interpretation is that contest-hooting was not merely expressing emotions, fear or hostility, but expressing them in such a way as to influence and change the behaviour of the other, avoiding conflict. Cognitive purpose and intention to control the relationship is also recognized here.

Bonobo vocal sounds are as purposive as those in the chimpanzee vocabulary but, in degrees of subtlety, bonobo vocalizing offers even more possibilities. These observations give important information; we have pointers to the social tactics that must flourish in bonobo groups in their natural habitat.

Mixed Signals: Bonobos and Humans

There have already been some studies of bonobos carried out in the wild. Observational studies were completed using audio-tape recording at a provisioning site at Wamba, Zaire (Mori, 1984). Mixed signals in bonobos' aggressive displays were a specific subject of the research analysis. Vocal sounds, described as defensive, were used during these aggressive displays. Bonobos' high-pitched hooting, for example, was a sound more commonly associated with flight but was used in aggressive displays together with a "defensive face". Unfortunately, Mori did not elaborate in detail, and leaves us with some interesting questions for future study. We may gain some insight into this mix of paradoxical signals, by considering the analogous mixed signals of human speech. Analyses of changes in human speech in competitive situations have emphasised the changes in pitch that sometimes happen (Schubert, 1991). Reactions to criticism, counterarguments, or responding to someone else's initiative, are expressed vocally at higher pitch levels. These are defensive speech reactions suggesting a parallel with the pitch changes in the excited contests of bonobos.

Comparing the hoot sound in chimpanzees and bonobos, the hoots of bonobos seemed less antagonistic in meaning. Yet one form of hoot, the hoot from a distance, is used both by chimpanzees and bonobos to keep groups together while travelling. The bonobo hoot from a distance was answered by "wraaah" (an alarm call) if the callers were strangers (Mori, 1983). We need more of this style of reporting on the dialogue aspects of signals.

SOCIAL PURPOSES AND VOCAL SOUNDS

The question is often asked whether vocalizing is semantic (referential), emotional, or both? I would prefer to ask: does the sound serve a social purpose, influencing others? In earlier research, a view of vocalizing as largely emotional was put forward in some of the classic studies. Goodall's (1986) list of descriptive labels for chimpanzee vocal sounds, refers almost exclusively to the expression of emotions—fear, anger, rage, distress, sexual excitement—but when she discussed each item, giving its typical usage, she referred continually to purpose. The chimpanzees wanted other individuals to do something or they wanted to prevent them from doing it. Animals such as dogs and cats act similarly but, for chimpanzees, the list is long and occasionally the items are complex.

Goodall's descriptions are still a valuable guide for research. For example:

1. "Hoo" sounds classified as part of the infant whimpering distress sequence were also used by adults in begging for food or for grooming.
2. The "cough-threat" soft bark made by higher-ranking towards lower-ranking individuals was a threat with a specific result, it worked to prevent a subordinate from moving closer or taking food.
3. Soft grunts, uttered when a pair of chimpanzees were feeding or travelling together, appeared to send a signal when one of the pair paused or got up to move on.

The descriptions were anecdotal reports but here is a promising area for closer study.

We have come some way now, in considering social control, from the view that vocalizing is an expression of spontaneous emotion, the *automatic triggering* point of view. The emotional component certainly produces no problems for the understanding of social tactics in humans. Human conversations have plans or purposes even though vocal changes in pitch, stress, and volume express emotions. The theoretical position of some linguists suggests that the purposive use of signals is precluded by the dominance of the limbic system and the centres of emotion in the brains of monkeys and apes (see Lieberman, 1995; Wallman, 1992). Thus, vocalizing will express emotions, dispositions, and behavioural inclinations such as flight or fight. According to Wallman, the received wisdom in primate studies was that the vocal calls of monkeys and apes were largely emotional in significance. This view of received wisdom is no longer supported. Research has moved on. In many social relationships vocalizing is representational, about food, predators, antagonists, or objects in the environment. Many vocal sounds are aimed to influence the actions of others: to prevent, change, or encourage certain actions in the listener.

CHAPTER SEVEN

Vocal signals: Monkeys

INTRODUCTION

Vocal signals have a variety of meanings and those that achieve some degree of social control, influencing other individuals, are especially of interest. From the social psychological perspective that I have followed here, we should be interested in this process of control as much as in the semantic qualities of the signals as precursors of language. In studies of vocalization, lists of sounds drawn up by researchers were the starting point. Despite some discrepancies, important conclusions were also reached, in early work, about the grading of sounds, so similar to human speech. Current research deals more with the dynamics of social actions.

Vocalizations now take a prime position in the research literature. In this chapter I give the historical background to contemporary studies as well as discussing recent work. In this way, the methodology and theoretical concerns of researchers and the advances they have made can be better understood. Historically, two field studies that produced comprehensive lists of baboon sounds were those of Hall and DeVore (1965) and Ransom (1981). Sound spectrograph studies have only recently been carried out with baboons despite the fact that baboons are very prominent in primate research (Cheney, Seyfarth, & Palombit, 1996). In the case of baboons' conversational or close-range grunts, it is possible that variations in the sounds may be signals with distinct social effects and this possibility will be considered in the light of the latest studies.

For savanna baboons, a basic list of sounds was drawn up by Ronald Hall and Irvin DeVore (1965). Later, Timothy Ransom (1981) undertook an 18-month observational study of olive baboons at the Gombe Stream National Park. Ransom drew up a list of signals that corresponded closely to the Hall and DeVore list, but included extra vocal sounds. Facial signals like tongue-protrusion and jaw-clapping that seemed to be important in close contacts, were also included. Ransom reported signals from a greater variety of individuals. It should be mentioned that Hall and DeVore were influenced by the dominance theory fashionable in the 1960s, which produced some emphasis on adult male aggressive signals with less attention to signals between females or between females and males (Fedigan, 1982).

Ransom, who followed the baboons on foot, was able to make a more detailed study than DeVore, who observed from a vehicle. More sounds were identified in the later study. There were more variations of grunts, divided into basic grunts, broken grunts, pant-grunts, and the hum-roar grunt. It was now noticed that soft grunts were frequently made, in friendly approaches, by both females and males. Finally, there was a possibility of dialects even between neighbouring troops. Plenty of such evidence exists for other species (Green, 1975, Mitani, Hasegawa, Gros-Louis, Marler, & Byrne, 1992). The baboons' vocal calls were individually distinguishable. There has been much evidence since then confirming the distinctive, individual characteristics of calls for both monkeys and chimpanzees. (Chapman & Weary, 1990; Hohmann & Fruth, 1995; Mitani & Brandt, 1994).

Detailed studies of baboons have, in the past, focused on movements and gestures, rather than on the vocal repertoire. But hints have been given of grading. In general, there is little more information available than in an outline ethogram. The most that could be said, without benefit of spectrographic analysis, was that basic grunts were a group contact vocalization or greeting; broken grunts were associated with initial aggression and blended into pant-grunting; pant-grunts occurred in an aggressive chase or attack; roar-grunts occurred in male displays and so on. As was the case in many early studies of baboons, researchers' interests did not extend to vocalizing but involved other aspects of behaviour, demographic patterns, and ecology.

RHESUS MONKEY VOCAL SOUNDS

Among the macaques, the most comprehensive lists of vocal signals exist for the rhesus monkey. The first study was Ray Carpenter's (1942) brief effort from his observations at Cayo Santiago Island, off Puerto Rico, followed by Michael Chance's (1956) fuller catalogue from a captivity study. In the 1960s, three independent studies appeared: one by Stuart Altmann at Cayo Santiago (see Appendix 1, p.145 for more on this location) in 1962; another was Robert Hinde and Thelma Rowell's studies of captive groups at Madingley, near Cambridge, in 1962, and Vernon Reynolds' study at Whipsnade Zoo also in England, completed in 1961. Only Hinde and Rowell used spectrograms to develop their analysis of graded, blended sounds. Table 7.1 shows the lists of vocalizations of three of the studies as summarized by Reynolds (the comparison was done

TABLE 7.1
Comparison of Rhesus Monkey Vocabularies

Reynolds (1961)	*Rowell & Hinde (1962)*	*Chance (1956)*
Hough	Bark	Hough-hough
Hough	Pant-threat (soft)	Hough-hough
Hough	Pant-threat (loud)	Hough-hough
Hough	Roar	Hough-hough
Hough	Growl	Hough-hough
Bark	Shrill bark	
	Squeak	
Gecker	Gecker	
Geckering screech	Geckering screech	
Screech	Screech	Eech
Scream	Scream	
	Caught noise	
Feeding hough	Food bark	
	Greeting growl	
	Long growl	
Girning	Girning (loud)	
Girning	Girning (soft)	
Explosive cough	Explosive cough	
Explosive cough	Movement call	
Explosive cough	Separation call	
Explosive cough	Conversation call	
Wailing		
	Shut-out call	
Food call	Food call	Coo Coo
	Baby calls	

From Reynolds, 1976, reprinted with permission.

before Altmann's study became available). It illustrates correspondences and differences in classifying. Using this comparison, we can see the discrepancies in overall numbers of distinct sounds.

As might be expected from the technical sophistication of the sound spectrograph, Rowell and Hinde's (1962) study had more descriptive categories than the other two studies and they also distinguished between movement, separation, and conversation calls. Altmann (1965) had compared his study with Hinde and Rowell's; he pointed to the disparities which meant that only six sounds out of many in each of the lists could be classified as equivalents.

The full impact of Reynolds' comparison became apparent in his final analysis. Here he revealed that, although there were 16 behaviour types common to all the studies from Chance (1956) onwards, only three of these were sounds (see Table 7.2). All the studies owed something to Chance's earlier list but, considered in relation to the total number of categories used by the researchers, the degree of agreement was not high.

TABLE 7.2
Consensus on Rhesus Monkey Vocal Sounds

Chance (1956)	*Reynolds (1961)*	*Rowell & Hinde (1962)*	*Altmann (1962)*
(1) Hough-hough	Hough	Bark, pant threat, roar, growl	'!Ho!'
(2) Coo Coo	Food call	Food call	'Koo'
(3) Eech	Screech	Screech	'eee'

From Reynolds, 1976, reprinted with permission.

One explanation of the differences was the varied social mix of the monkey groups such as presence or absence of mothers, infants, or juveniles. This might have affected the results. No doubt the different interests of researchers contributed. These were the "under the carpet" biases. Why, for example, did Altmann split maternal and sexual behaviour into so many more subunits than other researchers? Altmann gave no special rationale for this. One reason for the differences could have been the researchers' backgrounds in anthropology or zoology or other disciplines, but what were their implicit social theories, the hidden items mentioned by Reynolds? The comparisons of these studies had drawn attention to bias in the selection and interpretation of subject matter but they had also provided a careful and open scrutiny of results. Finally, the comparisons confirmed that studies made before the introduction of the sound spectrograph had been superseded.

CLASSIC PROBLEMS IN STUDYING VOCAL SOUNDS

Spectrographic studies now made available a precise analysis even if there were still many problems in interpreting the meaning of the signals. How can we identify changes in sound structure with changes in meaning for the hearer? How do we interpret the social meanings of subtle changes?

Complex sound structures are made more difficult for researchers to interpret because there are not only auditory characteristics to take into account but also situational cues. Categorizing, therefore, has to combine several different sources of information. A playback experiment has shown that young rhesus monkeys vary their screams depending on whether they are threatened by a higher- or lower-ranking individual or by a relative. The screams differed by tone and modulation; they were *noisy*, *pulsed*, *undulated* and so on (Gouzoules, Gouzoules, & Marler, 1984). In this experiment, at Cayo Santiago, the mothers responded, by looking towards the loudspeakers, more to the noisy screams than to their offsprings' other screams. Noisy screams were uttered against a higher-ranking opponent and in situations with physical attacks. Evidently, they referred to the type of opponent and severity of the threat. Screams conveyed urgency or danger present, for example, in an attack from a high-ranking nonrelative. It is still tempting to suggest that the sheer degree of terror could have produced such differences but the researchers denied this possibility because the screams were distinctive rather than forming a graded series. The study draws attention to subtle differences in sound structure which could carry a social message.

CHENEY AND SEYFARTH'S RESEARCH ON SEMANTIC SIGNALS

Major advances both in methods and explanations were made by a series of field experiments in Africa. Researchers studied the distinctive predator warning calls of vervet monkeys: for snakes, humans, leopards, or eagles. It was a process of discovery. Initially, two of these warning calls appeared to alert other monkeys to minor predators; a third to be used as a warning against snakes and humans. A fourth call seemed to be a warning after sighting leopards, lions, and eagles. Peter Marler (1977a, p.88) had commented that the signals could be considered as "functioning symbolically . . . designating particular referents." He had searched for links between the sound and its referent, snakes or leopards, but, at that stage of research, links could not be confirmed in relation to any *one* reference object. The calls seemed to have a mixed content; they were both an emotional response and a signal to warn others. This was no different from many human communications in which emotional arousal and cognitive purpose mingle (Marler, 1977a, 1992)

Dorothy Cheney and Robert Seyfarth (1980, 1990a) reappraised Thomas Struhsaker's (1967) research on these calls and Marler's (1977b) interpretation of them as signals "naming" predators. Critics had argued that they might all be fear-calls, reflecting only different degrees of fear. This was a possibility because the calls were loudest for leopards, less loud for eagles, and least loud for snakes. In a playback experiment, the calls were now played from a tape recorder. If the calls expressed only varying degrees of fright, the reaction to them should have been the same in each case—to scatter and run. On the other hand, if the calls carried different meanings, the vervets would stay in trees for a snake call and jump into a tree if it was a leopard call. What actually happened? The hearers first looked in the direction of the sounds and some ran into trees after a leopard call, others looked up or ran into bushes after an eagle call. If it was a snake call, some reared up, looking across the ground.

Was it Only a Fear-Call?

Other evidence supporting the fear-call argument was the observation that when an eagle actually swooped down, male monkeys sometimes gave leopard calls, but this may have been as a last moment urgent "scatter!" call. The leopard call was also noted on a very few occasions when the vervets under observation met another group; it was made mostly by one male and when the group was on the defensive. It was probably used as a diversionary stratagem (there are many reports of signal falsification in birds to distract competitors for food: Munn, 1986). Again, it may have been a fear-call with more emotional than referential content; some researchers have argued that vocal signals can be classified on a continuum of varying emotional or referential content (Gouzoules, Gouzoules, & Ashley, 1995; Macedonia & Evans, 1993). Purposive and emotional elements are both present in many human sounds. In the pre-linguistic vocal sounds of the newborn child, crying is an involuntary response under limbic/hypothalamic control but, as the infant matures, increasing cortical control produces crying on *purpose* and *to get attention*, a different sound structure recognized by the parents (Lester & Boukydis, 1992).

The playback experiments had shown that vervet monkeys could "name" predators and that their distinctive vocal signals alerted others to the nature of the danger. They did not call when they were on their own (and neither do birds). On the other hand, as Cheney and Seyfarth (1985) demonstrated, vervets give more alarm calls when in the company of vulnerable others, especially their offspring (and so do birds).

There are many species of birds, including starlings and chickens, that have different predator warning calls. Many birds have at least two, one for predators on the ground, another for those in the air (Cheney & Seyfarth, 1985). Birds also have special abilities that primates do not possess, such as their incredible and baffling ability to navigate; feats of memory that far surpass those of primates are carried out by nutcracker birds in retrieving stores of seeds at multiple locations even after a lapse of months (Balda & Kamil, 1988). For these reasons, there is no point in trying to argue for a linear scale of abilities with primates at the top of the scale. At the same time, however, the claim that monkeys and apes have the capacity to give vocal signals with symbolic content is more interesting because of their multiple relatedness to humans in neurophysiology, brain structure, anatomy, and genetic composition. This gives the study of nonhuman primates its particular fascination.

Conversational Grunts: Vervet Monkeys

Having achieved these results Cheney and Seyfarth (1982, 1990b) next explored topics that are more directly of social psychological interest. They looked for meanings of vocal signals in other contexts, recording grunts made between monkeys in close proximity while foraging, resting, grooming, or playing. The grunts sounded the same to the human ear but they evidently carried different meanings for the monkeys. Struhsaker (1967) had identified four situations in which grunts were used by vervet monkeys: in approaching a dominant; in approaching a subordinate; watching other monkeys moving or themselves initiating a group movement; when catching sight of another group. Although grunts were used in these different social situations, they could not be clearly distinguished by spectrogram analysis and there was no apparent difference in reactions to the grunts, except for changes in the direction of gaze.

The problem was whether the grunts were simply expressing feelings, or reflecting events in the environment. Were there acoustic differences that had been overlooked? A field experiment using a concealed loudspeaker gave some answers to these questions. *Grunts to a dominant* for instance, produced a change in direction of gaze towards the loudspeaker, whereas *grunts to another group* caused changes in direction of gaze towards the horizon. The grunts differed acoustically and had distinctive meanings for their hearers. The study had made an important advance in understanding the subtle differences in close-range sounds. Some of the sounds evidently had social effects: in particular, they were useful in controlling relationships relative to dominants or subordinates and in coordinating the actions of the group. Immature individuals needed time, about two years, to learn the situational, social uses of these grunts. Understanding what the sounds represented was an adult skill (Cheney & Seyfarth, 1986).

Other research efforts have been useful to a social psychological approach. Kim A. Bauers (1993), for instance, identified 26 vocal sounds of stumptailed macaques that were distinguishable by human listeners, in a one-year descriptive study. Bauers reported that one of these sounds, the staccato grunt, was almost invariably used in the context of a female showing friendly interest in an infant. Of 366 staccato grunts that she recorded, 95% were in this context. Interest in infants was also shown by line of regard, inclination of the body, and attempts at touching. This study emphasized the calming (and controlling) effects on the receiver of these signals, the infant's mother, suggesting also that the sounds had a purposive aim, clarifying the intentions of the signaller.

SIMILARITIES TO HUMAN SPEECH

Staccato grunts were close to the average duration of human speech syllables; they were used at close range and similar in intensity to human conversational speech. Some juveniles apparently had not yet learned to make these sounds and if they approached an infant they were threatened by the mother. This study (Bauers, 1993) gave more information about close range grunts and their uses than had previously been available. They were a signal of benign intentions in the particular context of approaching a mother-and-infant rather than a general appeasement signal.

An acoustic analysis of close-range calls between geladas (gelada baboons) has suggested that the sounds are similar to vowels and consonants in human speech, and also similar to speech in their rhythmic qualities. They seemed to carry messages between social partners both while grooming and while they were apart, feeding (Richman, 1976, 1987). Research also provides descriptions of squirrel monkeys' vocal exchanges at close range. These sounds seem to carry distinctive meanings (Biben & Symmes, 1986). A large amount of work has been done on the acoustic properties of New World monkey calls (Snowdon, 1982; Snowdon & Cleveland, 1984; Symmes & Biben, 1992) and this now needs complementing by attention to the social meanings of the sounds.

We have to search around for information on the social psychology of monkeys, sometimes from experiments where the researchers' interests are not directly oriented towards this kind of information, nor could we expect them to be. Often, research has a functional emphasis, as in showing that vocalizing influences spatial coordination during foraging. Field studies of squirrel monkeys and capuchin monkeys have established this (Boinski, 1991, 1992, 1993; Boinski & Mitchell, 1995). It is an important advance but now further questions persist about the caller's intention as well as the meaning for the hearer. Intentions are very much under conjecture here and likely answers have to be sought in the relationship between caller and hearer (receiver) of the calls.

Captive Japanese monkeys, just before grooming sessions, exchange low-amplitude signals similar to a coo. It seems as if these are an invitation and its acceptance. A playback experiment and spectrographic analysis have given new evidence of this (Masataka, 1989). Variations in call structure clearly depended on whether the caller wanted to initiate the session as groomer or was accepting the invitation. The value of this study came from its attention to proximate motives.

AWARENESS OF AUDIENCE: SUPPRESSION OF SOUNDS

Voluntary control over vocalizing suggests brain activity in the neo-cortex, the area of the primate brain most recent in evolutionary development, as compared with the involuntary reactions that derive from the more ancient mammalian brain areas in the limbic system (Lester & Boukydis, 1992). In this aspect of neo-cortical direction, vocal sounds in nonhuman primates seem once again similar to human speech.

There has been no shortage of references to the suppression of vocal sounds because of anticipated consequences. This behaviour is partly functional (where safety or survival is involved) but may also be of interest in the proximate context if it reflects tactical thinking about other individuals. A striking example of voluntary control over vocalizing was shown in the sudden silences of chimpanzees travelling in the forest when they were near hostile chimpanzee troops (Goodall, 1986). There are also some well known cases, in baboons and chimpanzees, of suppression of vocal sounds by sexual partners when an alpha male is nearby (Kummer, 1967). Goodall described how an adolescent chimpanzee learned to suppress his food calls when he was given bananas because on previous occasions larger males, after hearing his call, had come and taken them away. Spider monkeys diminish their food calls if the number of arrivals at a fruit tree is large. These monkey calls also appeared to vary depending on the abundance of food (Chapman & Lefebre, 1990).

THE SOCIAL MEANING OF VOCAL SOUNDS

Spectrogram analysis has showed that vocal repertoires of monkeys have been underestimated. One of Cheney and Seyfarth's (1990b) experiments has thrown light on an aspect of social behaviour that, in its subtlety, is an excellent example of a proximate social mechanism—signalling an individual's acceptance into membership of a social group. Structural differences between grunts (which sounded very similar to human ears) enabled the researchers to specify the period between the arrival of vervet males into a new group and their acceptance into that group *socially* as distinct from their *physical* presence. Although the males had joined the group physically it took 12–15 days until the resident members' *grunts to another group* changed to *grunts to a dominant* or *grunts to a subordinate*. In this way, acceptance of the newcomers was signalled, the message had a social meaning for the group.

At the social psychological level, studies of vocalizing are giving us evidence of the purposive uses of sounds especially in the conversational or close-range context; a variety of grunt sounds carry information about (or influence) events and relationships, perhaps involving some aspect of control over others' behaviour. These signals are especially interesting because of their cooperative uses. Zoologists emphasize the distinction between two kinds of signals: noncooperative signals (which are preponderantly the signals studied by ethologists) and cooperative signals (Krebs, 1991; Krebs & Dawkins, 1984). Noncooperative signals are loud, conspicuous, and stereotyped. Cooperative signals are inconspicuous and low-amplitude, likely to be overlooked by human observers.

Conversational Vocalizing

Even though there is now an encouraging number of reported cases of the semantic and controlling uses of vocalizing, we still need more information on conversational contexts. A search for evidence of dialogues yields few examples, but only very recently have there been studies of this type of signal in nonhuman primates. They have already shown new-found differences in meaning between graded but acoustically similar sounds. The research situation has been transformed by evidence of "new" sounds that previously went unrecognized. Cheney and Seyfarth (1990b), for example, had speculated and later gave empirical confirmation that vervet monkeys' conversational or close-contact grunts, sounding the same to a human listener, in fact had different meanings for the monkeys.

Naming of social events is one semantic use of vocalizing that should be of particular interest to us. What social events appear to be important ones? Telling others about the approach of another group or about an approaching stranger, signalling acceptance of a newcomer, greeting an approaching dominant: all these are important social events for each of which a distinctive vocal sound has evolved. Other vocal sounds are intended to control social relationships, with cooperative purposes such as calming an anxious mother of an infant, avoiding a fight, or coordinating movements in a group. For social psychology this is the way ahead.

The low-amplitude grunts, coos, or trill of female chacma baboons have been studied in the wild, in Botswana (Cheney, Seyfarth, & Silk, 1995a). This species has linear dominance hierarchies and it is well known that higher-rank individuals often vocalize when approaching a subordinate. The sounds seem to have a reassuring effect especially when used by females approaching mothers with infants. They are also the vocal equivalents of the positive gestures of touching, hugging, and grooming that research studies by de Waal (1989a), Aureli (1992), and Cords (1993a) have shown to be effective in reconciling former opponents.

Sixty percent of grunts were made to mothers of young infants, and eighty five percent of friendly interactions between former opponents included a grunt from the former aggressor. An experiment was carried out using the playback of vocal calls from nearby locations. A researcher followed a dominant (A) for 10 minutes after she had threatened or chased an unrelated subordinate (B). Some of these dominants had subsequently grunted, some had not—and this provided the experimental condition—reconciliation or no-reconciliation. Now, within 30 minutes the researchers played a recording of A's threatening screams to B, hypothesising that B would react more strongly if there had not been a reconciling grunt after the original incident. The hearers of the grunts in these two conditions behaved as predicted: the *after reconciliation* hearers reacted less strongly (they looked less towards the direction of the sounds).

THEORETICAL DISCUSSION

Useful as this work was, it was surpassed in a subsequent study by the same researchers. This time, the theoretical discussion makes an appreciable advance because of their efforts to assess both functional (evolutionary) and proximate (psychological) aspects in a field study of the loud barks of female chacma baboons (Cheney et al., 1996). The study is notable for the consideration that the authors give to two aspects of the calls: their function (monitoring contact between members of the group during foraging?) and their proximate causes (state of the signaller, intention to inform others?). The calls are very loud, can be heard at 2Km distance, and could be distinguished spectographically and by ear. The authors point to a problem that is common to all studies of long-range calls: although hearers may use the calls to maintain contact or to reach food, the caller may not have had this intention (Chapman & Lefebre, 1990; Smith, 1977).

One question was whether these calls were selectively answered by others. The problem here lay in the paucity of answering calls. They might have been answering calls of close relatives but this could not be statistically confirmed. In studying human social groups, in social psychology, counting the frequencies of communication is a well established method leading to understanding of, say, leadership or problem solving in a group. Frequencies of calls and answering calls might have held the key to their meaning. There were some indications of selective answering between close kin but the answers came primarily when those answering were in the rear and separated, suggesting that these were still contact calls, made mainly with reference to location.

The fact that females always answered a bark with the same call (and never gave such barks in response to any other call) meant that they probably interpreted the call as related to the context of separation. Some barks were made when the caller herself was not separated from the group and the proximate cause may have been separation from particular individuals rather than from the group. Here was a hint that a closer focus on these social relationships might yield understanding of the purposes of the calls.

We have seen some examples, in this chapter, of the analysis of monkeys' vocal signals in their social context. In some classic cases, researchers have succeeded in decoding the meanings of sounds but, in the true conversational contexts of dialogue this is proving to be difficult. It involves studying social interactions in which the nature of the communication is subtle. Clearly, this is a more demanding task than was the decoding of the more emphatic signals of threats and conflicts.

In the next chapter I begin a discussion of social tactics, devices intended to influence others. Studying the nature of vocal and gestural communication also involves a process of discovery about the social stratagems of primates. Negotiating and evaluating a likely opponent, attempting a challenge, recognizing rank, reconciling, forming alliances or consortships, are all activities that come into a social psychological focus.

CHAPTER EIGHT

Tactics and Social Devices

INTRODUCTION

In this chapter I give five case studies of the use of social tactics, actions gaining an advantage, in competitive and cooperative relationships: monkeys and apes acting with or against others to achieve an objective. The first case is that of coalitions or alliances between male baboons. The second is the rather baffling tactic of infant-buffering among male Barbary macaques and baboons. The third focuses on the tactics of female Barbary macaques in borrowing infants, also known as aunting or allomothering. The fourth is taken from Frans de Waal's (1982) incomparable account of the long-term strategies of male chimpanzees in a dominance contest. Lastly, I discuss the tactics of reconciling, when former opponents make and receive friendly gestures, shortly after conflict. Recent discussions of reconciliation highlight the use of specific gestures—the kiss of chimpanzees, the clasping gesture of Tonkean macaques, in a sequence of gestures that seem to be coordinated specifically to making their reconciling purpose more clear to their receivers (de Waal & Aureli, 1996). The topic of reconciliation is attracting discussion from a larger number of researchers compared to the four earlier cases when the scientific community involved in research on social behaviour was relatively smaller.

CASE 1: COALITIONS BETWEEN MALE BABOONS

In this type of coalition between male baboons, the two males attack a rival male which has a female partner. The two begin with a similar attacking role but then take different roles; one following through with threats against the male, the other guiding the female away. There is coordination of roles between them. Information on exactly how the threats, feints, attacks, and changes of role take place was not reported in the original classic study by Craig Packer (1977). The study focused on outcomes; which of the two enlisted an ally, which gained the female, and whether there was evidence of alternation of roles. The theme was whether the coalitions expressed reciprocal altruism, affecting reproductive success. The subject of the study was not the anticipatory actions of the pair or the way in which these were coordinated. However, for any progress in the current understanding of baboons' tactics, a social psychological focus of this kind will be needed.

The study raised questions about the tactics of the two males in the coalition. It raised the possibility of a pattern of reciprocal risk-taking and of alternation of the two roles, winner and helper. But what if there was asymmetry in the outcomes? This issue was taken up in a series of studies of baboon male coalitions in the wild (Nöe, 1989, 1990, 1992; Nöe, van Schaik, & van Hooff, 1991) with particular attention to the distribution of rewards between the partners. It might require the weaker partner in the coalition to wait over long periods before he was winner rather than helper. Extremely asymmetrical partnerships were, in fact, observed to endure for up to a year. Over 200 hours of observation taken over a period of a year, showed that three males joined in 93 coalitions; 28 were in conflicts over a female with one beneficiary only, 53 were also against high-ranking males with other immediate causes (Nöe, 1989). Might the asymmetry in coalitions over females be compensated for in coalitions of other types? Observations indicated that compensation was unlikely to restore the balance. Undoubtedly both coalition partners did benefit but perhaps in different currencies: one in terms of reproductive success, the other in proximate benefits such as protection in fights with other males. Strong asymmetry in gaining access to females did not mean that one partner received no benefits at all. Such questions touch on important general issues of primate social intelligence. Monitoring of coalitionary benefits involving the memory of past outcomes, of credits and debits, would imply a high level of mental ability (Tomasello & Call, 1994).

These speculations remain for future investigation, but Ronald Nöe thought that important clues to the relationship lay in the signalling between the partners. In his studies he had introduced a closer focus on the signals exchanged between the two male partners and he asked (1989, p.92) whether these signals might "make clear to the partner what otherwise only might have become clear to him after a long period of collaboration?" Nöe was evidently intrigued by the meaning of the signals (ibid., p.51):

> While the existence of true negotiation is hard to demonstrate, it is certainly possible to recognize behavioural patterns of baboons that could play a role in direct negotiations . . . coalition partners frequently communicated before and during coalitions and their communication seemed more intense when coalitions were more risky.

In these baboon coalitions, which supplant a rival male from consortship with a female, only one member of the coalition takes over the consortship. In Packer's study, the basis of the coalition seemed to be, at least potentially, a tit-for-tat strategy (each partner returning the favour by alternating in the role of winner or helper). Head-flagging by one partner initiated the coalition and he was the one to obtain the female. In this way, by alternation of the initiating appeal, each partner would have his share of the outcomes. Other researchers in the 1980s reported that the initiator of a coalition was no more likely to be the beneficiary than was the joiner. One study offered support for the notion that the strategy was cooperative: each coalitionary intervention yielded unequal costs and benefits but these balanced out over a period of a year and a half. Head-flagging was done by both prospective partners and each stood an equal chance of success at the start of their coalition (Bercovitch, 1988).

The three researchers, Packer, Nöe, and Bercovitch, reached divergent conclusions but there were very different conditions for each study. Moreover, the actors in the Packer and Bercovitch studies were olive baboons whereas Noe studied yellow baboons, and each study was at a different location with groups of varying size and composition.

CASE 2: INFANTS AS BUFFERS

Infant-buffering refers to males carrying an infant when under threat from another male. This particular action has often been observed by different investigators (Dunbar, 1988; Packer, 1980). One classic study of Barbary macaques has led the way in explanations of infant buffering, coining the term *agonistic buffering* (Deag & Crook, 1971). This term updated Itani's (1959) use of the term *passport* and the interesting suggestion that males carried infants as *passports*.

In their field study of Barbary macaques in North Africa, the researchers saw males frequently picking up baby macaques and approaching other males with them. The infant might be only a week old but the mother rarely objected. This action was usually done by the subordinate of two males. The two males exchanged a variety of signals, lip-smacking, teeth-chattering, touching, presenting, and mounting. A full explanation for these actions was not offered, only that in some way they allowed subordinate males to stabilize their relationship to dominants, to stay near to them without risking an attack.

Later studies gave new clues for the explanation (Smuts, 1985, 1987). There were two kinds of evidence. First, infants were carried towards higher-ranking and probably threatening males. Second, carriers of infants were less likely to be threatened, less likely to retreat, more likely to supplant others. The carrier repeatedly used a particular infant for this purpose. Males carried the infants that they normally protected and the infants ran to them readily. The likely benefits to an infant from its relationship to the male were protection and access to better feeding sites. The benefits of being carried appeared to outweigh the risks.

Infant-buffering also brought in the infant's mother as an ally. If this happened, there was actually a quadropartite relationship—male, protégé, ally, and aggressor. By the evolutionary logic, the least-committed male (without the infant) should be the one least willing to take risks. The payoff for the male with infant (survival of himself with offspring) is greater than the payoff of his rival (approach to a female).

Other discussions of infant-buffering have emphasized the genetic relatedness between the male and the infant. In one study of chacma baboons, it was argued that males carried their own infants to protect them against other males (Busse & Hamilton, 1981). As is well confirmed, infants are often killed by males that have recently entered a troop and were not present when the infant was conceived. The carrying males were longer-term residents in the troop and their rivals nearly always new to the troop. Evidence from other studies has only partially supported this hypothesis and the debate continues.

SOCIAL PSYCHOLOGICAL EXPLANATIONS

In social psychology, would we try to explain things in the same way? The answer is surely no. The explanation would focus on the thinking of the individuals concerned and their attribution of motives to others. If such a proximate focus for theory had been adopted for baboons, obviously we would know more about their social tactics. On the other hand, these baboon social interactions, despite their subtle communicative signals, are basic and direct, related visibly to reproductive competition. Both Nöe and Smuts use and refer to rich sources of information in the immediate context of communication and signalling, and these have highlighted the skills of baboons as social tacticians.

CASE 3: ALLOMOTHERING OR AUNTING

Another important topic in the social behaviour of Barbary macaques was the handling of infants by females other than the mother (Deag & Crook, 1971). This was referred to as *female care*. The persistent interest of *aunts* or nearby females in newborns and infants has been observed in many primates (the term "aunting" was coined by Rowell, Hinde, & Spencer-Booth, 1964). To approach the infant, female adolescents or juveniles often use the stratagem of grooming the infant's mother. The advantage for the mother is that she gains more time for obtaining food, and the infant may now have an additional protector. Young females can practise carrying, caring for, an infant. There can also be disadvantages. If the aunt is clumsy, the infant may suffer. A higher-ranking female may refuse to return a starving infant. From a broad discussion of macaques, baboons, and langur monkeys (*Presbytis* species) comes the conclusion (Hrdy, 1981, p.99) that females sometimes borrowed infants to use them "as pawns in social interactions or as props to practice maternal skills."

Allomothering has posed a problem for evolutionary theory (Nicolson, 1987). It has been difficult to relate it to reproductive benefits. The notion of *mothering practice* was first suggested by Lancaster (1975). In species with status hierarchies, such as macaques and vervet monkeys, it is quite likely that the aunt gains the infant's mother as an ally. In New World white-faced capuchin monkeys (*Cebus capucinus*) in the wild, allomothering is very frequent and involves allonursing (Perry, 1996). When mothers are separated from their infants during foraging it is very advantageous for infants to be able to nurse from the closest lactating female.

In many species, a female may be higher-ranking but she will not easily be allowed to approach an infant unless she expresses friendly signals. In the aunting overtures to Barbary macaque mothers, the approach is very gradual and tentative, females seem wary of frightening the infant (Zeller, 1991). The tactics are more complex than meets the eye. The individual approaching a mother-with-infant may have an ulterior motive that does not directly involve the infant, such as grooming or being groomed by the mother. Approaching the infant could be just a stratagem to secure the mother as an ally. Even more complex, the mother-with-infant could be used as a social means to develop a grooming partnership with another female close by. These possibilities illustrate the skilful devices that monkeys use.

I illustrate one such approach now. The female, approaching a mother-and-infant, has a flexible repertoire of signals to choose from. She may try out different signals in a sequence of actions, as shown in Fig. 8.1. This figure shows a sequence reported from field observations of olive baboons (Smuts, unpublished data). It shows an ingratiating approach by the female Clea, evidently intending to act as an aunt to Psyche's infant or to groom and become an ally of Psyche.

Clea's gestures, rhythmic grunts, play invitation to the infant, and gentle touching, are sufficient for her to approach to the point when she begins to groom Psyche. Grooming as a instrumental social tactic takes the foreground in this situation. Approaching must be accompanied by friendly facial expressions and vocal sounds. Lip-smacking, teeth-chattering, and quick glances are among the signals used; if the female approaching is of higher rank she has to overcome the nervousness of the mother. Her actions are controlled, matched to the mother's actions.

CASE 4: CHIMPANZEE TACTICS

During Frans de Waal's long-term observational studies (1978, 1982, 1992), he described a dominance struggle between three adult chimpanzee males at Arnhem zoo. Dominance was defined especially by the "rapid oh oh", pant-grunting sounds from a lower-ranking to a higher-ranking chimpanzee. Challenges to a leader took the form of "bluff displays" with rhythmic stamping of the feet. By monitoring these and other signals, it was known exactly when the dominance

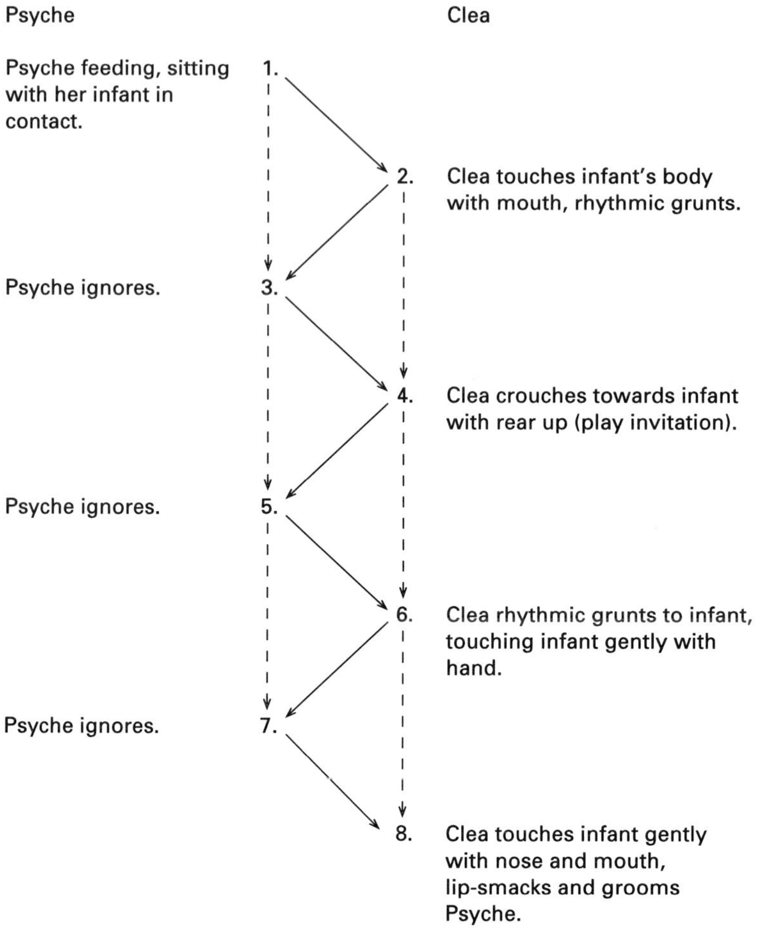

FIG. 8.1. Psyche and Clea.

order changed. Changes in alliances coincided with this change. The period between the first challenge to the leader in the form of a *bluff-display* by the challenger, Luit, and the leader's (Yerouen's) eventual displacement, signalled by his *rapid oh oh*, lasted two months (see Table 8.1). The chimpanzees' actions were extremely consistent. The third male, Nikkie, was the challenger's ally and he continued in this role subsequently. In the first month after the dominance change-over, Nikkie intervened on 31 occasions in conflicts between the new leader and others, and 30 of these interventions were in the new leader's favour.

Nikkie also intervened in a more complex and indirect way than merely by supporting the challenger in fights. He did so by attacking the female coalition partners of the leader. This tactic later proved to be of considerable help to Nikkie in his own rise in dominance and eventually in his take-over of leadership from

TABLE 8.1
Dominance Changes between Three Male Chimpanzees

		Dominance Positions		
		Yerouen	*Luit*	*Nikkie*
1st period	4 months	1	2	3
2nd period	2 months	1	2	3
3rd period	2 months	Change-over	Change-over	Change-over
4th period	1 month	3	1	2
5th period	3 months	3	1	2
6th period	3 months	3	1	2

Adapted from de Waal, 1978.

Luit at a later period. Table 8.1 shows the time-periods for one dominance change during the study.

Another development between the three males occurred in the period subsequent to the change-over. This was marked by Luit's new strategy, after his victory, of supporting the displaced leader against the likely next challenger, the younger male Nikkie. Competition now developed between Luit and Nikkie for the support of Yerouen. The competition again confirmed the purposeful nature of the chimpanzees' tactics and the capacity of each to negotiate with the two other individuals.

SOCIAL PSYCHOLOGY OF TACTICS

I have discussed social tactics in infant-buffering, aunting, and the complex realpolitik of several primate species. The most sophisticated examples were the anticipatory and devious actions of chimpanzees (de Waal, 1992). Some of the tactics described have been baffling in their indirectness particularly when one individual used another as a means to achieving a goal in relation to a third individual or even in relation to a third *and* fourth individuals. In infant-buffering, a male baboon used the infant as a means to influencing a third, the infant's mother, to act against a fourth individual, a male opponent (Smuts, 1985).

The various interpretations of infant-buffering have illustrated to what extent different researchers can offer alternative explanations. But it was the descriptions of signals, by facial expressions, by gestures, by vocalizing, and touch that give the basis for their inferences. In the case of chimpanzees we saw evidence of sustained tactics which came to form longer-term strategies. In de Waal's (1982) interpretations of the social events many concepts are used that are familiar to social psychological analysis including "respect" and "attitude". Yerouen's attitude towards Nikkie, for instance is rated for its fluctuations –1 to +1 over a two-year period. This may seem a somewhat rash tranference of a mentalistic concept from humans to chimpanzees but actually the index is based on Yerouen's pro or con actions, a perfectly valid basis for inferring attitudes, in chimpanzees as in humans.

CASE 5: RECONCILING GESTURES

Reconciliation, identified as friendly approaches between former opponents immediately after conflict (de Waal & van Roosmalen, 1979), has been the topic of many recent studies, mainly in captivity, of monkey species but including also chimpanzees, bonobos, and gorillas. Usually, a former victim approaches an aggressor with gestures such as embracing, kissing (in chimpanzees), touching with the hand, grooming, and vocalizing (for instance, grunting) with facial signals such as lip-smacking or teeth-chattering. Among the currently accepted explanations of such reconciliations are that they repair damaged social relationships (Cords, 1992; Cords & Aureli, 1996), signal acceptance of the dominance hierarchy (de Waal, 1986b), and reduce stress on the loser (Aureli & van Schaik, 1991).

Among many other issues, there is the methodological question of how long individuals should be observed to be sure that a reconciliation occurs. Usually, the first few minutes after a conflict are considered crucial. Limiting the period of observation to brief time periods also leads to curtailment of other sources of information, hampering our understanding of social relationships and structures, important influences on reconciliation (Kappeler & van Schaik, 1992). But this situation is already changing as new research studies are reported.

Current research is exploring aspects of reconciliation that are of interest to a social psychology of primates. Signals, gestural, tactile, and vocal, that are specifically used in reconciliation are being identified, as Cheney, Seyfarth, and Silk (1995b) have done for baboon vocalizing. It is also recognized that the history of interactions between partners may affect whether reconciliations occur (Cords & Aureli, 1993) and this is a promising pointer for further research.

Most recent studies of reconciliation have been undertaken in monkeys in captivity. The analysis has also been extended to wild gorillas (*Gorilla gorilla berengei*). In this species, reconciling occurs by brushing against, touching on the back, embracing, and (here, a glimpse of the full reconciliation process) *grumbling* (Watts, 1995a). The social relationships implicated in reconciliations await future study, especially the connection between dominance styles and reconciliations. In species with a rigid hierarchy, there may be less reconciling (Butovskaya, Kozintsev, & Welker, 1996; Matsumura, 1996) or within a network of alliances more individuals may be involved (Watts, 1995b). In pigtailed monkeys, reconciliations took place between aggressors and the kin of victims and between aggressors and their own kin, possibly as reassurances, reducing tensions (Judge, 1991).

The reconciliations observed in chimpanzees and macaques can be considered as a window on primate cognition (de Waal & Aureli, 1996). A second category of post-conflict behaviour—consolation—has been identified in chimpanzees. This is "succourant behaviour" (de Waal & van Roosmalen, 1979) and occurs when a participant in a conflict afterwards seeks reassurance from an uninvolved third individual. Similar kinds of behaviour have been reported for gorillas (Watts, 1995b) but not for macaque species. This difference may reflect emotional and cognitive capacities present in apes but not in monkeys (de Waal & Aureli, 1996).

Two hypotheses may account for how this difference may have appeared. The first, the *social constraints hypothesis*, states that reconciliation and consolation are found only in species with high levels of cooperation and attachment (see also Matsumura, 1996). The rigidity of a dominance hierarchy may also be a central factor. The absence of food-sharing in macaques, outside the mother–offspring relationship, sharply contrasts with the widespread sharing of food among chimpanzees in their more flexible social relationships.

The second, the *social cognition hypothesis*, states that the difference occurs because of the higher mental capacities of apes, suggesting empathy in the case of consolation. Empathy is usually associated, in humans, with the emergence of a sense of *self and others*. In the context of consolation, there is a wide repertoire of chimpanzee signals: the pout, whimper, yelp, outstretched hand, or shaking with both hands, as well as screaming. Macaques have fewer and much less dramatic displays. Chimpanzees thus seem to show a much greater responsiveness to the emotional states of others. This behaviour corroborates the significant research studies on chimpanzees' attribution to others of knowledge, feelings, and intentions (Povinelli et al. 1990), tending to suggest chimpanzees' greater capacity for perspective-taking in relation to others.

In the next chapter, I consider some puzzles in the reciprocal actions of primates, examining the consequences of social grooming, an important feature in the relationships of monkeys and apes. There may also be an opportunity for theoretical adventure. Can we use a social psychological theory to explain these actions?

CHAPTER NINE

Social Exchange and Grooming Partnerships

SOCIAL EXCHANGE IS NOT RECIPROCAL ALTRUISM

When a social psychological focus is applied to nonhuman primates we can ask: What are the immediate social motives, aims? How much thinking about tactics takes place? Recent studies of social interactions raise the possibility that monkeys and apes exchange favours in many situations. Is grooming a form of social exchange? Before we continue with these questions there is a matter of definitions to sort out. The difference between social exchange and reciprocal altruism has to be made clear, because the two concepts can mistakenly appear as identical (Cosmides & Tooby, 1993): "Sometimes known as 'reciprocal altruism', social exchange is an 'I'll scratch your back if you scratch mine' principle". This error can be put right, simply by adding that reciprocal altruism is the *evolutionary version* of social exchange.

Stating that social exchange is the same as reciprocal altruism brings confusion because of the ultimate and reproductive consequences that we assume for reciprocal altruism (Trivers, 1971) and because reciprocal altruism incurs risk of physical detriment whereas social exchange does not. The confusion in the Cosmides and Tooby (1989) definition came about because they took social exchange to mean the general class of reciprocal relationships, including reciprocal altruism as a special case. I prefer to keep the concepts separate, recognizing the distinctive theoretical level of each—proximate (the immediate situation) for social exchange and ultimate (having reproductive consequences) for reciprocal altruism. The reciprocity in social exchange is altogether different from that of reciprocal altruism. In the latter, the timescale for benefits is more remote and expressed in evolutionary and genetic payoffs whereas social exchange treats with relatively short-term, proximate rewards and costs.

Confounding phenotype with genotype occurred even in the classic discussion of reciprocal altruism by Trivers (1971). Certainly, warning cries in birds, or rescue from drowning in humans, should be understood in evolutionary terms as reciprocal altruism. But in Trivers' discussion of food or implement-sharing in humans, he moved the theoretical level to the proximate context of social exchange, without signalling the shift. It was, however, noticed and commented on by social psychologists (Myers, 1983; Sahakian, 1982). Altruism, which should be explained in the currency of reproductive gains and losses, came to be confused with cooperative behaviour in general. As Callan (1984, p.413) mentioned, an analytical construct referring to the genotypic had been transposed "to the phenotypic classification of actions and emotions. The illicit move is greatly facilitated for Trivers and others by failure to keep control over the twists and turns of metaphoric language . . ."

SOCIAL EXCHANGE IS NOT EXPLOITATION NOR IS IT ALTRUISM

Having recognized the difference between social exchange and reciprocal altruism, the question now is whether monkeys and apes engage in exchange. In human groups, if exchanges of rewards (actions or objects) are to continue, rewards must outweigh costs. Many social relationships can be understood as providing rewards but certainly not all do so; in some there is little or no exchange. For instance, exploitative relationships are scarcely reciprocal, nor is there a social exchange where self-sacrifice and risk reflect altruism. As mentioned, reciprocal altruism differs from social exchange because it includes risks of physical detriment and the payoffs are judged to be in the form of reproductive success: either as an individual's genetic advantage or as an increase in relatives' reproductive success (inclusive fitness). In reciprocal altruism, social actions are assumed to be linked with numbers of surviving offspring or second-generation descendants (Dunbar, 1988). Social exchange, on the other hand, involves exchanges of actions or objects for which a direct biological linkage to genetic fitness need not be claimed either at the theoretical or empirical levels.

From a functional viewpoint, benefits and costs of reciprocal altruism should be measured in terms of individual survival and reproduction (Harcourt & de Waal, 1992). Zoologists, in general, have been interested in reciprocal altruism and reproductive success, and not in social exchange at the proximate level. Their research on social relationships has considered reproductive advantages. Female alliances are understood to protect the partners from harassment by dominants, give better access to food, and consequently improve birth-rates. Male alliances improve reproductive success, as estimated in increased frequency of copulations. Tracking the reproductive effects of coalitions has been a constant interest.

A few researchers have given attention to social actions without seeking direct links to biological outcomes. Some have placed social events and actions at the centre of their research programmes, seeking explanations at the proximate level, while referring to ultimate theory as background. By giving primacy to social events and their immediate consequences they had taken what was, a few years ago, an unorthodox position in primatology. In a discussion of the tactics of chimpanzees (de Waal, 1986b, p.460), there is an explicit statement of this position:

> Throughout, my emphasis will fall on *proximate* explanations—those based on immediate causes and goals—and on the social histories, experience and intelligence of the animals involved. This treatment . . . will differ from the currently predominant genetic/evolutionary perspective.

THE USES OF GROOMING

Now I turn to cooperative behaviour where partners exchange mutual benefits at little cost to themselves. The case I have in mind is that of social grooming, an activity that takes up a great deal of the time of many monkey species but uses little energy. By all the evidence, grooming is relaxing and pleasurable for the groomee and seems often to be competed over, as a resource. There are frequent accounts of monkeys being displaced from grooming pairs by more dominant ones.

In captivity studies of rhesus monkeys, it has been estimated that monkeys spend 27–37% of their total time in grooming (Reynolds, 1981) although all social activities increase in captivity because there is no foraging or travelling. From field studies, baboon and macaque species are estimated to spend 15–18% of their daily time in social grooming (Dunbar, 1993). In a group of free-ranging stumptailed macaques, during a period of 100 hours, joining others and grooming others were the most frequent actions (Estrada, Estrada, & Ervin, 1977). Conflicts, huddling with others, displacing others, presenting and mounting, contributed relatively less. In general, grooming is thought to be the most common form of social activity in nonhuman primates (Muroyama, 1991, 1996).

Many investigators have followed J.B. Watson's (1908) lead in interpreting grooming as serving some important social function over and above its hygienic functions. Several authors have interpreted grooming as bond-forming and bond-maintaining. This connection is unclear because the term social bond covers a variety of activities (Goosen, 1980). Some researchers understood grooming as the expression of a social bond rather than its cause (Washburn & DeVore, 1963). Nevertheless, there is wide acceptance that grooming and affiliation are closely associated (Cords, 1995). In many monkey species, resident and related females groom each other more than migrating males.

Numerous studies have confirmed that grooming assists social integration and that it has calming effects. In situations of likely tension the frequency of grooming increases, as, for example, between male langur monkeys during females' mating periods (McKenna, 1982). Grooming increased between male/female pairs of hamadryas baboons on the appearance of a male rival (Kummer et al., 1974). In chimpanzees there is frequent grooming between males, a tendency that is associated with dominance conflicts and the forming of alliances.

Instrumental Uses and Investment in Grooming

The immediate instrumental uses of grooming have been much emphasized. Grooming allows juvenile male rhesus monkeys to approach dominant males and be accepted by the dominant into a social unit (Simonds, 1973). A similar instrumental use of grooming was reported in female rhesus monkeys wanting to hold an infant; they first approached and groomed the infant's mother (Zeller, 1991). There are various uses of grooming as a tactic: a female chimpanzee was seen to groom two males in an effort to reconcile them (Goodall, 1968). Grooming also sends a message to others in the vicinity: a male gelada A, on being threatened by male B, rushed to male C and groomed him, an action clearly telling B that A and C could be allies (Dunbar, 1984).

Grooming may involve longer-term investment. Although we have no firm causal evidence of this, there are some plausible indicators. Female baboons grooming males receive advantages such as a share of food and space, and the grooming partner defends them against other males and females (Smuts, 1985). Grooming might be an action communicating a promise of future commitment (Simpson, 1991). A male gorilla, by grooming another may be cultivating an ally (Harcourt, 1979). An economic analogy, of the kind found in social exchange theories, considers grooming as an investment for acquiring future support (McFarland, 1987).

If grooming is offered by females approaching a mother-and-infant is it an exchange or an instrumental tactic? Does the female initiate grooming as a tactic to get close, perhaps to gain the mother as an ally, or is grooming exchanged for allomothering? In the latter case, which of the two benefits most? In allomothering, the mother is relieved from carrying the infant and gains more time for feeding. The benefits for the substitute mother involve mothering practice or play. An observational study of patas monkeys showed that substitute mothers invested vastly more time in grooming than mothers. It could be because the balance of benefits tended more towards them than to the mothers. This is consistent with the hypothesis of a social exchange (Muroyama, 1994). It could also be that grooming the mother is done, instrumentally, to approach the infant; the substitute mother has learned that grooming the mother has the result of bringing her close to the infant and so the tactic is due to a lower level of learning rather than to transactional behaviour.

The concept of a balanced exchange has also been used to explain reciprocal grooming among Japanese monkeys. Between kin there is no necessity for returning a benefit promptly, but unrelated pairs have to do this to maintain the relationship. The notion of an exchange of favours is confirmed (Muroyama, 1991). In marmosets (*Callithrix jacchus*), in the wild, breeding females were central in grooming networks. There is a strong tendency for these females also to groom nonbreeding ones, suggesting an exchange of grooming for help in caring for offspring (Digby, 1995).

Grooming, Fight Interventions, and Altruism

Could the end result of grooming a dominant, plausibly be an alliance? Dominant individuals are groomed more than lower-rank ones and lower-rank individuals compete to groom them. The classic studies by Seyfarth (1977, 1983), showed that high-ranking female monkeys were the most sought-after as grooming partners but, because of the constraints of competition, middle- and lower-ranking females tended to settle for partners at their equivalent level of rank. Cords (1997, p.36) also points out that, in the species studied, "rank distance is correlated with kinship, so grooming adjacently ranked animals might alternatively be explained by the attractiveness of relatives." In this case, the social implications of grooming are unclear.

There are also further concerns. Grooming could be an item in a bargain between groomer and protector but there are theoretical difficulties to face. The stage might appear to be set for us to find evidence of an exchange between grooming on one side and acting as an ally on the other. This would be a convincing scenario were it not for the risks of physical injury (which would be more consistent with altruistic actions). Conceptually, the notion of a social exchange is plausible only if the opponent is unlikely to offer resistance, an intervention without risk of injury. Risky behaviour is better understood functionally as reciprocal altruism involving reproductive fitness. The concept of a social exchange, as I mentioned earlier, does not entail life-threatening costs, it is simply an exchange of proximal benefits.

In the different case of tit-for-tat actions in helping a partner against adversaries, the explanation is relatively straightforward: this is reciprocal altruism, with risk-taking by both partners. However, if an intervention is made on behalf of a grooming partner, the action is one-sided and the theoretical rationale must be that the action is altruistic. The protected individual is, doubtless, a close relative. The partner intervening in the fight, protecting the groomer, takes all the risk, while the groomer has expended relatively little (only time and small amounts of energy). This is why it is most likely that acting in fights on behalf of a grooming partner is influenced by kin relationships. The empirical evidence bears this out (Harcourt & Stewart, 1989; Kaplan, 1978).

There is little likelihood of an action against a higher-rank opponent on behalf of a nonrelative. Coalitions with kin can be explained by the theory of inclusive fitness and, in most species (except for chimpanzees) non-kin coalitions are a small minority of the totals observed. Moreover, in chimpanzees the non-kin coalitions are expressed in reciprocal interventions and not in terms of grooming as the item exchanged (de Waal & Luttrell, 1988). Across the cercopithecinae (macaques and baboons) individuals intervening in defensive coalitions do so mostly on behalf of kin and undertake greater risks than when they act on behalf of nonrelatives (Fairbanks, 1980; Silk, 1982). These studies refer to tit-for-tat actions in contests (an intervention now by one partner being "repaid" by an intervention by the other partner), rather than to any supposed transaction between grooming and such actions.

A comparison study of fighting in captive groups of chimpanzees, rhesus monkeys, and stumptailed macaques, studied reciprocal-altruistic interventions, testing the hypothesis that they were based on the cognitive ability to keep mental records of given and received behavior (de Waal & Luttrell, 1988). In a correlational analysis (after partialling out the effects of proximity, kinship, and sex), residual correlations showed some tendency to reciprocal interventions in macaques but more so in chimpanzees. Studies of monkeys' fight interventions, in general, have not confirmed a reciprocity rule and statistical artifacts make results unclear (Hemelrijk, 1990a, 1990b; Hemelrijk & Ek, 1991).

Fighting on the side of a grooming partner may certainly be altruistic and reciprocal (if the partner intervenes at another time) but it is not reciprocal altruism merely because of the presumed return favour of *grooming*. There is not sufficient detriment or risk to the groomer to make his or her actions altruistic. Grooming does not "break through the threshold" required for it to involve a reproductive cost (Dunbar & Sharman, 1984). The energy costs of grooming are trivial and it is hard to see how this behaviour could be altruistic.

Researchers have come to similar conclusions about the trivial costs of grooming. For baboons and brown howler monkeys, the cost of grooming in lost feeding time is small (Chiarello, 1995; Saunders & Hausfater, 1988). The hypothesis that grooming and being groomed lead to reduced vigilance against predators was not borne out in an observational study of wild blue monkeys *Cercopithecus mitis stuhlmanni* (Cords, 1995). A captivity study of rhesus monkeys did, however, suggest there were costs in the form of harassment of an infant while its mother's attention was distracted by being groomed (Maestripieri, 1993). It is not known if this holds for monkeys in the wild.

Conclusions. The Costs of Grooming

The critical evaluation of concepts has centred on the question of whether mutual grooming should be considered in the same explanatory framework as fighting for a grooming partner with risk of injury. The answer, in my opinion, is that it should not. Grooming is to be understood at the proximate level only; it does not break through to reproductive effects. Support in fighting, on the other hand, is in many cases directly related to reproductive success: either, in males, by increasing opportunities to copulate or, in females, protecting them and their infants from harassment.

GROOMING AND SOCIAL EXCHANGE

If we discount that grooming is altruistic or reciprocally altruistic we are now free to discuss grooming as an item in social exchange: as in mutual grooming or, more interestingly, grooming for other benefits. Is it likely that grooming is an action that either anticipates or is made in return for other favours? Researchers have been pessimistic about this, not accepting that nonhuman primates are able to make comparisons across time or to have the cognitive capacity for exchange over the long term. Chimpanzees' capacity to anticipate wants in the future had been estimated as a matter of minutes or hours. On the other hand, chimpanzees recalled actions of other chimpanzees over periods of days or longer and acted on them later, in cases of retaliation (de Waal, 1992).

Evidence for grooming as an activity traded in exchange for benefits comes from four field studies, two of chimpanzees, one of bonobos, and another of talapoin monkeys. In these cases the trade-off between grooming and its reward is more or less immediate. A convincing description has been given of a dominant male chimpanzee being groomed by another male before the groomer was allowed to approach a female; here grooming was the price in a process of bargaining (de Waal, 1982). When a young chimpanzee met a dominant male sitting in his way on a branch, he groomed the dominant for a short time before stepping over him, a kind of quid pro quo. (Sugiyama, 1969). If the males were of similar size, one simply stepped over the other. In the case of a dominant finding a smaller male in his path, the smaller showed submission by crouching, or withdrew. Among talapoin monkeys, grooming seemed to be a reward for particular actions. In the wild, males were seen to guide groups of juveniles over difficult crossings; the male was later groomed by them (Rowell, 1973). Bonobos appear to exchange grooming for food: they groomed food possessors giving them access to food (Kuroda, 1984). This example touches on the problem of interpreting what happened. There may have been an exchange but it could be that grooming was only used as a tactic, to get to the food (see also de Waal, 1989b).

In these few reported incidents in field studies, the return favour is also immediate. There are many obstacles to researchers' establishing a connection if there is delay between the first action and its return favour, if it occurs. Connecting an action to its antecedent in situations of multiple social events is a demanding task, yet the examples of mutual account-keeping in chimpanzees extend over long periods (de Waal, 1992).

An Experiment on Grooming and Interventions

A well known field experiment tested the hypothesis that grooming was traded for support in aggressive disputes (Cheney & Seyfarth, 1986, 1990a; Seyfarth & Cheney, 1984, 1988). The experiment involved playing vervet monkeys' threat grunts from a concealed loudspeaker. These grunts usually accompany actions against an opponent and other individuals often join the caller. During the experiment, the monkeys were observed as they participated in grooming pairs. After grooming finished and one of the monkeys had gone out of sight, her threat grunts were played to the remaining grooming partner. Unrelated individuals looked much longer *after grooming* with the vocalizer than when *no grooming* had taken place before the calls. Monkeys related to each other looked for about the same amount of time whether there had been prior grooming or not.

The interpretation of these observations has produced some unresolved issues. Looking towards the sound is not the same as giving support in an aggressive coalition. Was looking influenced by the consideration "What has she done for me lately?". It could be that vervets' attention to the calls might have shown they wanted to intervene but we cannot be sure. They may have looked because of a recency effect, merely recognizing the caller as the grooming partner. Perhaps, as Harvey Gouzoules (1992, p.158) argued, "a missing control condition—that *any* recent interaction, even a fight, might influence the tendency to attend to a call—is needed before the more cognitively complex interpretation can be entertained." Gouzoules added that no study has shown that monkeys are sensitive to the costs and benefits of assisting unrelated individuals, though Cheney and Seyfarth had come closest to doing so.

A new experiment is sometimes hailed as a break-through on a given problem; later discussions in the scientific community suggest alternative explanations. This pattern has been shown over and over in social psychology where variables are difficult to define without confounding errors (Cooper & Fazio, 1984).

An Experiment on Social Exchange?

An experiment with a captive group of long-tailed or Java monkeys possibly supports a social exchange explanation. The aim was to change the dominance relationships in the group. Monkeys previously low in dominance were trained to operate three levers at a dispenser delivering food pellets. Normally, dominant monkeys would get to food first, displacing less dominant ones. After introducing the newly trained "specialist" lever-pullers, any changes were recorded in the behaviour of the 55 pairs of monkeys, each consisting of a lever-puller and a dominant partner. The results showed that in 14 pairs out of 55, former dominants stayed close or followed the other; in 13 the lever-puller now received more grooming from the partner. Of these, in seven pairs, the lever-puller was both followed and groomed more (Stammbach, 1988).

Existing dominance relationships formed over the years in this group of Java monkeys probably operated against any profound changes. The problem is reminiscent of experiments in social psychology where the volunteers bring with them social attitudes and characteristics that the experimenter tries to control by random assignment to roles. All in all, the experiment is suggestive of a social exchange in the cases where grooming of the lever-puller increased. Alternatively, these changes could be explained as a result of conditioning, associating the presence of lever-puller with more food. There is some uncertainty in the interpretation.

Conclusions on Social Exchange

A hopeful, if unexpected, extension of the notion of social exchange emerges from a review of the sexual behaviour of chimpanzees and bonobos. Exchange in this case occurs in social interactions in which a female obtains nonreproductive benefits whereas the male obtains a mating opportunity (Wrangham, 1993). The exchange of sexual favours for food had earlier been reported among female bonobos (Kuroda, 1984). Grooming, food, and protection from aggressors are among the benefits the females obtain from this form of socio-sexual exchange.

A number of reports have shown that social exchange can take place among nonhuman primates. Chimpanzees, with their account-keeping of favours rendered, may be an exceptional case but it is one that encourages us to think that there will be more empirical support, in the future, for the voluntary, even calculative, choices of a true social exchange. As things stand, it certainly looks as if chimpanzees, and possibly monkeys, have some forms of social exchange. The more elaborate development of such activities in humans is an evolutionary attainment, but related tendencies exist in parallel primate lines.

The discussion of social exchange in contemporary publications is hardly a focused affair, nevertheless the proximal context of exchanges is of interest to many researchers and it is relevant to our increasing knowledge of nonhuman primate mental abilities. William McGrew (1992), on the basis of his long-term studies of chimpanzees, has raised a number of fascinating questions that await our attention. He suggests, for instance, (p.223) that:

> studies of reciprocity should widen their scope to include the *signals* used to initiate, maintain and terminate exchanges, as in food-sharing . . . We cannot now say how much chimpanzees can *negotiate*, such as "How many minutes of social grooming must I invest before you will let me use your hammer-stone?" . . . We need to know how good chimpanzees are as accountants.

The current empirical support for exchange in social behaviour is strongest for chimpanzees and less for other species of apes and monkeys. Yet there are promising leads to follow up in future research. It would be unwise to underemphasize the importance of this basic form of social behaviour across nonhuman primate species.

CHAPTER TEN

Sexuality in Monkeys and Apes

INTRODUCTION

How should we approach the study of sexual behaviour in monkeys and apes? It is not necessarily productive to seek similarities or differences. The seeker risks getting lost among the myriad small and large variations between primates. Instead, let us look for some general features of sexual behaviour from a social psychological perspective. The first striking feature is the easily visible nature of all sexual activities in nonhuman primates. This single characteristic contributes a huge distinctiveness. From a researcher's point of view it means all sexual behaviour is as open to view as any other kind of social event.

Second, for nonhuman primates, it is impossible to distinguish the social and sexual as separate areas of behaviour. Much of the social behaviour observed in apes and monkeys has immediate sexual objectives, leading to copulation. But a signal that appears to be sexual to a human observer, may actually carry a nonsexual message such as a greeting. The signal may be sexual in one situation but not in the next. Take the case of presenting; in one situation this is an invitation from a female to copulate, while in others presenting by both males and females is a friendly, appeasing, reassuring, or greeting signal.

In most contemporary human societies, there are rules for separating the social and the sexual, but for nonhuman primates there is no dividing line between them. When monkeys and apes communicate by touch, they may do so in ways that seem sexual although, in fact, the message is something quite different: for example, a baboon male's touching the genitals of another is often merely a reassuring signal (although it can be tactically very important). The contrast to human societies does not hold in all cases; for example, touching the genitals is a form of greeting between males in the non-European societies of New Guinea and Australia. Ken Burridge (personal communication) has described how an anthropologist, Siegfried Nadel, visited a student in the field in the Highlands of New Guinea: "On his arrival a man knelt before him, gathered Nadel's testicles in his hands and bent his head as though to kiss them. Nadel stood transfixed, horrified, unable to move. In the few years left him he refused to visit the Highlands again." Touching between males, even by a hand on an arm would be a powerful signal in our society.

There may also be quite commonplace situations, in human behaviour, that can be described justifiably as socio-sexual, as in the case of flirtation in initial heterosexual encounters when sidelong glances, head tilts, and leaning forward are all signals contributing to the sexualization of a developing relationship (Simpson, Gangestad, & Nations, 1996).

SEXUAL STRATAGEMS

If the objective of an animal's actions is copulation, then for many zoologists this is the significant fact. But the actions leading to copulation are themselves of interest to social psychologists because they form complex social tactics. A male chimpanzee, for example, negotiated with another male until he was allowed to approach the latter's female partner. The bargaining gambit involved was the amount of grooming received by the other male before he acquiesced (de Waal, 1982). There are many incidents of this kind in primatologists' reports. Sometimes copulation is not the objective but only a means to obtaining something else. One field worker noticed that among bonobos some copulations were initiated by females; their partner was a male with fruit in his possession. It suggested a transaction or exchange (Kano, 1989).

Mating systems in nonhuman primates provide the structural context that we need to take into account if we are to advance our understanding. The phylogenetic characteristics of each species, its physiological and anatomical features, constitute another set of givens providing the framework for socio-sexual communication.

Despite exceptions, anyone who is looking for human parallels can make a choice between two categories of primates: those that have frequent but sporadic sexual activity and those with more continuous activity. The former, for example chimpanzees, have multiple short-term mates. The latter, for example gibbons, have one long-term mate (see Appendix 1).

There are also two further categories which can be used for the purpose of discussion at a general level. These are the species that have extreme sexual dimorphism (with marked differences in size between male and female) and those that do not. This categorization tends to coincide either with polygynous (in the case of extreme dimorphism) or monogamous mating (in the case of slight dimorphism). So here we have biological features that are the background for the varied social behaviour of monkeys and apes. Let us look first at the categories of sporadic/continuous sexual activities.

Frequency of Sexual Activities

Mating in primate species either follows a pattern of sexual activity that is frequent and fairly continuous or seasonal and sporadic. The general mammalian pattern is definitely the sporadic one. But primates are different. In general, there is greater flexibility in their sexual behaviour. References to oestrus or heat traditionally emphasize physiological, physical, and behavioural changes during the period around ovulation (Beach, 1976, Hrdy & Whitten, 1987). During this period females are approached by males and they show invitational behaviour to males vocally or by gesture and posture. Although in primates there may be an increase in matings at this time, the tendency can vary over a wide range from the very evident to the barely noticeable. Martin (1990, p.437) considered it inappropriate to refer to oestrus for monkeys, apes, and humans because "female receptivity, though often maximal around the time of ovulation, is not tightly restricted to a specific stage of the non-pregnant cycle and is not so heavily dependent on hormonal factors" as in mammals generally.

An error of earlier studies was to stress continual sexual activity, which actually was the result of captivity conditions and the absence of environmental influences such as fluctuations in food supply. From the 1950s onwards, field studies began to report mating and non-mating seasons. Nevertheless, the idea of definite patterns of activity, associated with ovulation, oversimplifies the actual patterns that vary from species to species, even from female to female (Fedigan, 1982).

It would be misleading to consider sexual activity all-important, even though sexual relationships are often a key feature of social interactions. The social units of primates do not exist only for breeding purposes; there are other functions, such as defence and protection of infants and adults, and if breeding is seasonal, mating is involved in a relatively small proportion of total activities (Bernstein, 1970). Although many primates mate during pregnancy, few do so during the lactation period. Food supply, temporal changes such as length of day and climate, are all influences on periodic mating. Bonobos are an exceptional case: females' perineal swellings last longer than those of any other species and, unlike chimpanzees, females copulate at any time outside the periods of this physical change. For the rest, the periodicity of sexual behaviour is evident because of the long intervals of reduced sexual activity during pregnancy and lactation: 18 to 24 months for baboons, four to five years for chimpanzees and gorillas.

To summarize, there are generally two limitations on sexual activities in nonhuman primates. The first is the seasonally determined cycle of mating of many species in the wild, related to food availability for offspring. The second is the long period that females spend in pregnancy and infant care.

What is the interest for social psychologists in these patterns? We should, of course, take note of Irwin Bernstein's view, just mentioned, that sexual activities could be of relatively minor importance in the average day of many primates.

This is evidently not true for bonobos. What about other species? In any case, how frequent is sexual behaviour during the periods when females are active? Female chimpanzees may copulate up to 15 times each hour. Of the copulations of adult males, 96% happen during the female's two-week peak period of increased activity. One female copulated 50 times during one day; this level of activity tended to coincide with social excitement when travelling parties met or when food was found (Goodall, 1986). For chimpanzee females in their most active period, there is an average of five to six copulations in early morning, dropping to two per hour at midday, rising slightly in the afternoon and decreasing to one per hour thereafter (Tutin, 1979). Barbary macaques in the wild have been reported mating with up to 10 males a day during their most active periods (Taub, 1980).

In contrast to this, in the monogamous New World cottontop tamarins (*Saguinus oedipus*), copulations occur once per hour in newly formed pairs and slightly more than once per day in well established pairs. Unlike most of the Old World monkeys, marmoset and tamarin females are continuously active sexually. The comparison here seems to be between almost frenzied bouts of sexual activity in polygynous groups, and less frequent but steady mating behaviour in monogamous pairs or one-male units. Although baboons show high rates of periodic sexual activity compared to tamarins, summarizing the number of copulations over a year, marmosets and tamarins are likely to have higher totals because they are active throughout the reproductive cycle (Snowdon, 1990).

Copulations in apes are brief. Chimpanzee copulations last 7–8 seconds on average as recorded in field and captivity studies. In a field study, bonobos' copulations lasted, on average, for 15 seconds in the wild, 20 seconds in captivity. Bonobo copulations last from a few seconds to 80 seconds (Kuroda, 1980). Gorillas' copulations average 96 seconds in field studies and 52 seconds in zoo studies. For orangutans, the durations are 10 minutes according to field observations and 15 minutes in captivity studies. One well known characteristic of chimpanzees is the size of the scrotum, it is relatively very large, accounting for 0.269% of body weight, compared to 0.017% in gorillas and 0.079% in humans (Schultz, 1938; Short, 1979). It could be explained as an evolutionary adaptation to the high frequency of copulations among chimpanzees (Goodall, 1986). This particular measure might suggest that humans are biologically further along the way to monogamy than to polygyny, an extremely controversial issue in itself (see Miller & Fishkin, 1997; Zeifman & Hazan, 1997).

Female Initiatives

There is great variation between species in sexual behaviour but the choice of a partner depends first and foremost on the willingness of the female to mate. This is usually the decisive influence. Females can avoid mating with certain males (Perloe, 1992), but they often take the initiative in courtship and seek out and approach males for mating (Bercovitch, 1995; Huffman, 1992; Small, 1993). Female Barbary macaques even give an extra emphasis to the action of choosing by leaving one consort to pair with another. A variety of actions by females lead eventually to copulation, such as approaching, following, grooming, vocalizing. This holds true for monkey species and for apes in most cases. Female monkeys use postures, gestures, and facial signals. Among the most well known of these are flicks of the tongue by tamarins and howler monkeys. Another is puckering of the lips by New World capuchin monkeys and Old World patas monkeys. Lip-smacking is a frequent courtship signal in baboons and rhesus monkeys. In some species of macaques, female vocal calls are invitational to males; in gelada baboons copulation calls usually precede the male's approach and presenting by the female (Dunbar, 1988). Human sexually invitational signals, from females, are conventionally restricted to eye-contact, flirtatious glances, head tilts, smiles, laughter, and body posture. Despite the potential of these signals, the prevailing cultural script in Western society on how a courtship should proceed is definitely discouraging of female initiatives. It almost always follows the following sequence: A man and woman meet. One of them (most often the man) asks the other out. Later on, the relationship develops and becomes exclusive, again with the man taking initiatives, in a pattern that is so general that even to mention it seems trite (Fillion, 1997; Holmberg & Veroff, 1996).

The tendency to female initiatives that Taub (1980) reported in Barbary macaques, was confirmed in a nine-month study of this species by Small (1990, p.268) who found that the females mated, as she said, "in the manner of polygynous males." Small was very much aware of the influence of the social attitudes of researchers in the interpretation of female mating behaviour: especially because of contemporary views that males in general are supposed to be more interested in sex than females. Small argued that due importance should be given to female assertiveness. Male possessive behaviour, as observed in some other monkeys and apes, did not operate in Barbary macaques. Climate and food availability produced synchronized conception; sexual swellings lasted several weeks in the equivalent of a feeding frenzy.

Female Advertising

Genital swellings are a distinctive characteristic of baboons, some macaques, mangabeys, chimpanzees, and bonobos. In other monkey species—vervets, guenons and all other apes—this is absent. Despite very thorough efforts to give an overall explanation of sexual swellings and their presence or absence in different social systems, it has been impossible to do so completely convincingly (Clutton-Brock & Harvey, 1976). Even within macaque species there are variations: the fact that many species living in multi-male groups do show swellings suggests the functional explanation that swellings might attract several males, enabling females to choose between them, but there are exceptions or anomalies.

In the rigid social system of the hamadryas one-male breeding unit, females do show swellings. However, mating choices even here are not absolutely restricted to one male. The unit leader's possessive tactics are often evaded by deception (Kummer, 1982).

Over 20 species "advertise" females' periods of highest sexual activity with perineal swellings (Hrdy & Whitten, 1987). Swellings are larger in chimpanzees. In macaques, the degree of swelling is slight to moderate, but there are colour changes to the sexual skin on the rear, flanks, and other parts of the body. In rhesus monkeys, the sexual skin is an area that extends from the rear, forward onto the abdomen and face, down to the thighs. Barbary macaques and pigtailed macaques have marked swellings similar to baboons but stumptailed macaques show little outward signs of ovulation.

Why should females advertise? Certainly, swellings are characteristic of many species that live in multi-male groups, where males compete with each other in mating. Of the 24 species of colobines or leaf-eating monkeys, only three have social groups containing several adult males and these are the only three in which females have sexual swellings. In bonobos, genital swelling continues through pregnancy and usually resumes about a year after giving birth. Thus, female bonobos have full or partial swellings, continuously (Badrian & Badrian, 1984). Incidentally, it has been noted that, in human females, sexual swellings would be "completely incompatible with bipedal locomotion", that is, with "the locomotor needs of an upright, striding biped" (Morbeck, Galloway, & Zihlman, 1997, p.111).

In some species, there are physical changes in males corresponding to females' period of increased sexual activity. The size of the testes enlarges in male rhesus monkeys and there are skin colour changes. In squirrel monkeys, males increase in weight, developing fatty shoulder pads at the beginning of females' annual period of activity. It is not known if environmental changes affect females and males directly, or whether one sex is affected directly and influences the other by social communication. It seems plausible, in this case, that female physiological changes affect males. Odour is also important as a sexual signal. Research has identified *pheromones*, hormone-related substances that attract males (Keverne, 1976). New World monkeys' use of scent-marking suggests that cues by odour can also signal female "receptivity" effectively. The use of this term should be accompanied by a caution about implications and misunderstandings that may surround it. Sexual receptivity, according to Beach (1976, p.125), was a term "distinguished equally by the ubiquity of its usage and the infrequency of its definition." He warned against its possible implication of passive females receiving aggressive males, which would misrepresent the mating sequence and give a biased view of female sexuality. Receptive reactions were defined by him as occurring specifically in the consummatory phase of the mating sequence, involving posture and other facilitating actions at that point, immediately before and during copulation. Beach (1976, p.130) also emphasized the nonhormonal influences, in primates, on whether or not "a particular female will cooperate with a particular male", mentioning especially inter-individual preferences.

I have now referred to the extreme variety in female sexual changes. This is an aspect of physiology that will affect the forms of socio-sexual behaviour in different species. Female genital displays are more elaborate in multi-male groups and this has led to three functional explanations: the *best male* theory, the *many male* theory, and the *obvious ovulation* theory. These are not our direct concern here but I mention them because of their interest as competing theories. The best male notion argues that the swellings function to encourage competition between males and so increase the likelihood of the best male fathering the offspring. The many male theory argues that the swellings increase the female's opportunities to mate with different males, also increasing the likelihood of protection by the males subsequently for both mother and infant. The obvious ovulation theory holds that the swellings serve to demonstrate the timing of ovulation, thereby increasing paternal involvement on the part of males (Hamilton, 1984; Smuts, 1987).

The best male, many male, and obvious ovulation explanations all seem plausible. Certainly the multi-male social system is implicated. It is quite striking that there is a relative absence of sexual swellings and of male displays in species with monogamous and one-male units. These sexual characteristics need to be noted, as background to the social behaviour we observe. For a social psychologist it has to be the given framework within which to go into a fuller study of interactions.

Friendships and Consortships

The relationship of the male–female pair is our next topic, giving more scope for the social psychological perspective. There are a variety of possible pairings, of which one is a *friendship*, as in olive baboons, that does not have an immediate connection with mating and a *consortship* during the periods of sexual activity. Female and male baboons often form friendships for weeks or longer; in chimpanzees, consortships can last for well over a month: although female baboons initiate the approach, males have to use social skills in the early steps towards forming a relationship. Field studies have given us a systematic analysis of friendships. In forming friendships, males were observed to follow and stay close to a female and to groom her frequently. These tactics had to be used over several weeks, even months, until the formation of a friendship was signalled by a reversal of this pattern—that is, the female now started to follow and stay close (Smuts, 1985, 1997).

Chimpanzee consortships involve less finesse and more physical coercion (Goodall, 1986). But some consortships are calm, relaxed. The female's swelling was a sexual bargaining point according to Goodall, and its immediate benefits were increased grooming by males, more frequent shares of food; its costs were occasional intimidation. Chimpanzee consortships are associated with the female's period of genital swelling and are therefore much briefer than the long-term friendships observed in baboons. In fact, all our evidence on friendship formation and maintenance comes from the field studies of baboons. Accounts of mutual support and exchanges of benefits that accrue from baboon friendships, although not yet quantified, have been a consistent theme in researchers' interpretations.

Male Initiatives

In chimpanzee, bonobo, and bonnet macaque (*Macaca radiata*) courtships, males usually take the initiative. It is interesting to check the success rate of chimpanzee courtships. In a five-year period of observational studies from 1977 to 1981, males were observed courting females at the peak period of genital swelling, 1475 times. Only on 4% of these occasions did the female fail to respond within one minute. In the different case of refusal by females who were not at the peak period of swelling, there was a strong likelihood of attack by the courting male (Goodall, 1986). Altogether, there are three mating patterns in chimpanzees (Tutin, 1979):

1. Opportunistic mating, as when a group of males encounter females;
2. Consortship, when an adult male and female stay together for a week or two;
3. Male possessive behaviour, when a dominant male tries to prevent other males from approaching a female.

The opportunistic and consortship categories were identified during an observational study over two years at the Mahale mountains, Tanzania. Most of the 800 copulations observed were opportunistic, but chimpanzees in consortships would not be so available to observation because they tended to disappear from view (Hasegawa, 1989). Females migrating into a group courted males by using signals such as the clipping of leaves with their teeth, another instance of female invitations.

All copulations by bonobos observed during a four-month period at Wamba in Zaire were opportunistic; adult males mated with females in the vicinity. Bonobo males are less competitive in mating than chimpanzees. Only 34 of 515 copulations were interrupted aggressively by another male; this would be more frequent in chimpanzees (Kano, 1989).

Bonobo Sexuality

One unique form of sexual activity among bonobos is mutual genital rubbing between females, lasting up to 20 seconds and, more rarely, up to one minute. This behaviour was usually seen among females at periods of genital swelling but females outside these periods also participated. Genital rubbing occurred especially at a food provisioning site when an individual who had obtained food was surrounded by other females making begging gestures: it was interpreted as partly a form of tension reduction but once again the tactical uses of sexual bargaining could provide an explanation (Kano, 1980).

In general, the sexual activities of monkeys and apes show differences and contrasts to human sexuality. Bonobos are an exception. The similarities are: the reduced importance of the menstrual cycle, variations in sexual positions, and face-to-face copulations (de Waal, 1989c). Researchers have been quick to point to the similarities in bonobo and some human copulatory positions: 66 out of 106 bonobo copulations reported in Takayoshi Kano's (1980) field study were in the dorso-ventral position, 37 in the ventro-ventral position.

So far in this chapter I have given examples of behaviour that in many cases is truly socio-sexual. Sexual objectives are often the motive for social tactics or social objectives are the motive for sexual initiatives. Bonobos are the prime example. Their sexual strategies (more so than chimpanzees) also show that they are highly competent social actors. The evidence from this source fits well with other sources of information on their gestures and dialogues that I discussed earlier.

An analysis of bonobos' precopulatory communication was carried out in a detailed captivity study (Savage-Rumbaugh & Wilkerson, 1978). The complete sequence was summarized as follows:

1. Prolonged mutual gaze.
2. Initial exchange of numerous signals, facial and vocal.
3. Further posture changes and gestures, communicating the body posture to be taken by each (at least 25 signals were identified).
4. Taking up the position or gesturing until this is done; the researchers referred to this as *agreement* or *disagreement*.

The equivalent chimpanzee overtures were described from one field study as consisting of four stages, reflecting a less complex exchange (McGinnis, 1979) although there may be many other variants that were not included in this study.

1. Male gestures.
2. Participants approach.
3. Female crouch-presents.
4. Male mounts.

In bonobos, so far as current studies suggest, there may be more reciprocal gestures and the signals are less stereotyped. In male chimpanzees' signals, only one, the arm stretch, is a manual gesture. Bonobos' gestures suggest a capacity to make representations of the partner's body position and of the position into which the partner should move. Even the sheer variety of these positions makes bonobos somewhat different from other species in which copulation is relatively perfunctory, and bonobos' tactics are not coercive as in chimpanzees. Although information in the form of actual signal-by-signal sequences was not supplied, the bonobos' use of mutual eye-contact was evidently an important channel for coordinating behaviour. Male bonobos did not continue if the partner avoided eye-contact. The wide-eyes stare face with raised brows was among several facial expressions identified. This differed from the glare of chimpanzees in that it was mutual and the lips were not compressed. The brow was raised as opposed to being lowered or tensed. As well as showing the variety of copulatory positions, this study also emphasized the bonobos' tendency for ventro-ventral copulation. Only rarely have gorillas and orangutans in the wild or in captivity been seen to use this position. Studies suggest that bonobos are unique, as far as the subtlety of socio-sexual behaviour is concerned, and we can hope for more details about their social tactics from future research on a scale equivalent to what is currently known about baboons and chimpanzees.

We have come to the end of this discussion of the social aspects of sexuality. In indicating the intricate involvement of sexual goals in social interactions, we have entered an area of frequent controversy. Up to the 1970s, the coercive aspects of male sexuality in hamadryas baboons and chimpanzees were over-emphasized as parallel models for making comparisons with humans. Researchers have long recognized that there are other, alternative sources available. Cases like that of the Barbary macaques, showing females controlling sexual relationships, have been a healthy antidote to those traditional attitudes, as is the recent work on bonobos. We also know from these studies that bonobo males and females have almost equal power, both socially and physically. Their mother–son alliances have an important impact on the relative power of males (Kuroda, 1989). Bonobo males are only a little heavier or stronger than females. They are only slightly sexually dimorphic. This brings us to our next topic: dimorphism and its implications for understanding sexual behaviour in nonhuman primates and, possibly, of gender roles in humans.

PRIMATE MATING SYSTEMS

In nonhuman primates, the ecological distribution of resources, the food supply, and predator pressures are the most likely influences on mating systems, although these are not themes we can pursue here. Current opinion suggests that ecological pressures set the scene for the spatial and social relationships of females, and stable structures emerged out of the interaction between female distribution and male mating strategies. Female behaviour is more directly adapted to ecological pressures because female fitness is limited by environmental changes such as food availability (Wrangham, 1980, 1987).

One further influential factor in primate mating systems is the degree of sexual dimorphism, the differences in physical size between males and females, which vary tremendously among primate species. It is important to know about the range of dimorphism; which species are most or least dimorphic. If the male is very much larger than the female, the possibilities are greater for physical coercion of females and the assertion of dominance. With some exceptions, the more dimorphy a species has, the more competitive are the males and multiple mating is the rule (Mitani, Gros-Louis, & Richards, 1996). This is the scenario for chimpanzees and baboons but it is not at all the case for gibbons and New World monkey species, in which sexual dimorphism is slight. The more physically similar are males and females (as in the case of gibbons) the more they tend to have monogamous mating.

Primate mating systems have often been referred to as source models for humans, supporting arguments in favour of either monogamy, polygamy, or promiscuity as natural conditions. They are also source models for human gender roles. By selecting particular species of primates, support can be found for directly opposing points of view, from a view of male dominance as the key element in primate evolution to a view of females as dominants or as full participants in evolution (Zihlman, 1997). Earlier studies of the various baboon species and of chimpanzees and gorillas tended to support the male dominance position because these are the species with marked dimorphism, with males much larger than females.

Sexual Dimorphism and Male Competition

If we look at male–female physical differences across primate species, we find a range of variations in weight, height, dentition, musculature, and strength. These variations are not distributed neatly between families of species. Possible associations of sexual dimorphism with mating behaviour, larger size with males' competitive mating, do not always appear as this hypothesis would predict. The connection between slight dimorphism and monogamous mating patterns is certainly suggested in the case of species such as gibbons or titi monkeys for which long-term pair relationships form the basic breeding unit. The picture is clouded for some other species, usually considered monogamous: in marmosets and tamarins, which show little dimorphism, units of two males with one female (polyandrous) *and* pairs, have been observed (Goldizen, 1987).

One well known exception to the association between extreme dimorphism and multiple-mating is found in an Old World species, de Brazza's monkeys (*Cercopithecus neglectus*), the males are much larger than females yet male/female relationships are sometimes monogamous. Gorillas and orangutan males are over twice the size of females. Baboons exhibit a similar size difference. These species are certainly not monogamous. Chimpanzee and bonobo males are heavier than females (estimated at 36% heavier for the former, 23% for the latter). Human sexual dimorphism estimates indicate that males are between 18 and 30% heavier than females (Ghiglieri, 1987, 1989; Passingham, 1982). Feminist writers saw this difference as relatively moderate (in comparison with gorillas, it is) and they played down the importance of dimorphism as a characteristic that, in humans, would make inequalities between the sexes inevitable (Fausto-Sterling, 1985).

Female Choice and Feminism

The feminist and sexual revolutions of the 1960s and 1970s were influential in the emergence of a literature that started to assign due evolutionary importance to females' choice of mates. Female initiatives have now become the subject for research studies of a range of species including rhesus and Japanese macaques (Huffman, 1992; Manson, 1995). According to feminist writers, too much attention had been given to extreme dimorphism in the great apes, permitting advocacy of a biased view of the natural origins of male domination. Insufficient attention had been given, for instance, to the part taken by males in the care of the young of New World and other species (Haraway, 1989).

A critique of notions of male dominance had emphasized that female primates are no less assertive, competitive, and political than males, although more subtle (Hrdy, 1981). Bias in terminology and theory was effectively criticized (Fedigan, 1982; Small, 1993). Much evidence has been provided of active female choice and of complex mating tactics that certain (but not all) earlier researchers had treated as exclusively male domains. Female initiatives in courtship and mating have by now been reported on too large a scale to allow a view of promiscuous males and sexually conservative females (Small, 1993). These ideas touch on topics that are essential to our understanding of the social interactions of nonhuman primates. They are part of a theoretical discussion continuing alongside the empirical studies while at the same time influencing their content. The value judgements and cultural predispositions of researchers are never unimportant. Although they are not as directly our concern here, they are still vital to a full awareness of our subject matter.

As a result of this critique, references to the "sexual receptivity" of females could no longer be seen as innocent, because they suggested passivity (Herschberger, 1970). The critique has also argued against merely giving females a more active role in an explanation that is only a counterpart of the previously fashionable male dominance theory. It was argued by some that primate studies had stressed genetic factors too much and underemphasized the social context (for example, by Sperling, 1991). Women primatologists had made a significant contribution in identifying the biases in their discipline (Gruen, 1996). Some had erred towards exaggeration, accusing primatology of being politics by other means, a discipline in which theories are value-laden (Haraway, 1989). While accepting arguments referring to the social construction of science, this cannot be taken to the extent of undermining scientific efforts. Even if topics and subject matter are influenced by fashionable social agendas, such fashions change; we should be optimistic.

CHAPTER ELEVEN

Dominance and Social Relationships

INTRODUCTION

From a social psychological perspective dominance relationships are of interest because of what they reveal about the social tactics of monkeys and apes. It should be emphasized that, for many primates, dominance over others is not a major feature of their social lives and there are differences among the primate species in this behaviour as in others. Researchers, historically, assumed that dominance had ultimate effects in greater reproductive success. Such effects are still an area for controversy (Bérard et al., 1993; Pereira, 1995; Smith, 1993). In general, the topic of dominance has been over-researched in the past and does not attract much attention at the moment. But there is no denying that dominance contests and alliances are a central aspect of social behaviour, for both males and females in some species.

Dominance has not been a negligible issue in primate studies. Higher or lower rank are as influential in social interactions as they are in human societies. When sociograms of a primate group show the heavy concentrations of grooming choices towards an alpha male this is the visible indication of one important consequence of dominance. There are many others. Grooming a dominant can lead to increased access to resources. In chimpanzees, the social tactics that involve dominance contests are complex and well researched, and justify our interest in the proximate effects of social rank.

A major objective of primate communication, whether by gesture or vocalizing, is control over events and individuals. Control, in monkeys and apes, is used either to improve mating and feeding opportunities or to anticipate and avoid the aggressive actions of others. Dominants can use control to take priority in feeding and mating, although this does not rule out counter-actions on the part of others. Threats and coercion are effective sources of power but there is reward power as well: that is, the dominants' positive contributions to alliances, protecting their allies. As with power in human relationships, dominance is difficult to measure in nonhuman primates, partly because dominance may fluctuate relative to changing social relationships, the challenges of others, and the interventions of allies (Harcourt, 1988).

Perhaps it will be useful therefore to consider dominance not as an unvarying characteristic of an individual but as a commodity to be traded and competed over. Submission and grooming of the dominant have their trade-offs in alliances, protection, and sharing of resources. If the dominant position can be wrested from an individual, then conflict is the stratagem that is followed. In some monkey species, matrilines that can pass on their dominance to future generations are passing it on as pseudo-property, the nearest that nonhuman primates can approach to the human notion of owning something.

DEFINITIONS

Definitions of dominance are based on observations of social interactions between pairs of individuals; one of them is the winner, the other the loser in any encounter. The classic test for dominance, in the laboratory, is to place a peanut or raisin in front of two monkeys and to see which of the two reaches for it first. In the wild and in open-range situations, displacement of an individual by another is usually sufficient to classify which is dominant. Simply, one individual approaches, the other promptly moves away. In vervet monkeys, for example, displacement of one monkey by another is a very frequent event (Cheney & Seyfarth, 1990a).

The bared-teeth display in macaques, and pant-grunts in chimpanzees, both from a subordinate to a dominant, are consistent indicators of dominance relationships more than any particular incident of approach or retreat (de Waal, 1986b; de Waal & Luttrell, 1985). The latter outcomes are changeable depending on the proximity of alliance partners but the visual and vocal signals are reliable indicators of the current dominance hierarchy.

If defined as the capacity, at a particular moment, alone or with others, to obtain priority in feeding or mating, dominance cannot be disregarded in studies of social interaction. But how widespread are dominance relationships across species? In studies of Old World species such as rhesus and vervet monkeys, dominants receive more grooming from others and are sought as grooming partners as a way to avoid harassment from them. (Seyfarth, 1977, 1981, 1983). Subordinate monkeys continually monitor the dominants' actions and avoid getting too near without first making submission signals such as a fear-grin or a brief presenting gesture. Although it is clear that dominance cannot be considered as a universal, it is an important influence on the social behaviour of

TABLE 11.1
Dominance as an Intervening Explanation

Independent influence	*Explanation*	*Dependent activities*
Physical size; grooming partners		A displaces B for food, space
Hormonal condition: oestrus in females, testosterone levels in males	**"DOMINANCE"**	
Maternal rank		B "Bared teeth" grins to A (submissive gesture)
Alliances		male possessive behaviour with females

Adapted from Hinde & Datta, 1981.

macaques, baboons, chimpanzees, and gorillas. On the other hand, in species such as black-and-white colobus monkeys (*Colobus vellerosus*), redtail monkeys (*Cercopithecus ascanius*), patas monkeys, blue monkeys, and Tonkean macaques, social interactions seem egalitarian.

In tree-living species, such as gibbons and howler monkeys, dominance tendencies are hardly noticeable; in many species, investigators have looked in vain for signs of dominance. It seems clear now that studies from the 1940s to the 1960s overemphasized dominance by males, as a key to explaining primate social activities. Since then, researchers have found that female dominance over females is much more stable and important than male dominance in the social systems of macaques such as rhesus monkeys (Hrdy, 1974). In baboons, subordinate females are excluded from resources by dominant females, although across nonhuman primate species in general there are wide variations in such tendencies (Rowell, 1974; Smuts, 1997).

It has long been recognized that dominance is not one unitary characteristic influencing an individual's actions across a number of contexts. Dominance does not lead, in a deterministic fashion, to complete control in fights or in mating (Gartlan, 1968). It is not a fixed individual attribute but is likely to be the result of coalitions and alliances (Harcourt, 1992). Dominance relationships in ground-living monkeys and apes *are* a marked feature of their social behaviour but often involve more than two individuals if friends and relatives are near.

If the term dominance is used only to describe which monkey obtains food or mates, it has not explained how dominance came to exist. There is a range of possible sources of dominance, as illustrated in Table 11.1.

Dominance relationships do not reflect such permanency as was once supposed. In chimpanzees, dominance rank may rotate among the adult males only over long intervals, but in other species changes happen much more quickly. For instance, female monkeys' dominance status changes during periods of increased sexual activity. A sexually active female will gain more access to mates, food, space, or favoured locations. This seems partly due to the alliances or consortships that she now forms with males, as has been shown in chimpanzees, baboons, rhesus monkeys, and pigtailed macaques. However, the intense competition for mates, food, or space, described in some captivity studies, does not exist in the wild simply because there is an abundance of food, no shortage of space, and frequent mating opportunities.

DOMINANCE HIERARCHIES

So far, I have referred to dominance as it is expressed in social behaviour. There is also the aspect of dominance relative to a hierarchy of ranks. Rank necessarily involves a hierarchy of positions and is therefore a characteristic of group structure. Here we encounter what Hinde (1997) defines as the dialectic between levels of explanation, where the dominance hierarchy is a major causal influence on social behaviour and at the same time is a structural product of that behaviour.

Hierarchy or rank order can be identified from investigators' reports of displacements, threats, or other actions considered to be indicators of dominance. Zoologists have identified a dominance hierarchy in terms of behaviour measures reflecting the direction of aggression in a downward line from the most dominant individual A, to B, then from B to C, from C to D and so on (Schjelderup-Ebbe, 1935). In captivity studies, some primates have exhibited a clear hierarchical order. Although observers report hierarchies, this does not prove that they also exist in the minds of monkeys. Controversy has flourished over whether such hierarchies are perceived by the monkeys or apes themselves. It could be that each monkey knows only which of its immediate neighbours has to be given priority and which does not. Is a hierarchy anything more than a construction by human observers from the frequency of individuals' displacements of others? There is some evidence that monkeys are capable of transitive inference; that is, they can show awareness of a linear order of ranks.

In a well known study of the displacement of lower-ranking monkeys by higher-ranking ones, this awareness was strikingly illustrated (Cheney & Seyfarth, 1990a). When a high-ranking vervet monkey female approached two lower-ranking ones in a grooming partnership, one of the latter invariably moved away. Almost always it was the lower-ranking one. If a female ranked number 2 approached females of ranks 4 and 5, it was number 5 that moved. Therefore, monkey number 5 must have known that (a) number 2 was superior in rank to her and to number 4, and (b) that number 4 was superior to her. To that extent, she recognized a hierarchy of rank. The researchers had shown, very effectively, that the dominance hierarchy was definitely implicated in the social cognition of vervet monkeys and that this feature of group structure was influential in their social interactions.

WEAKENING THE DOMINANCE CONCEPT

There are contrasts in dominance behaviour between caged animals and undisturbed groups in the wild; the former are much more likely to have a rigid hierarchy. If monkeys are too closely confined, the result has been to exaggerate restrictive hierarchies. In the wild, dominance hierarchies are fluid because of the challenges of younger rivals as well as the opposing coalitions of lower-ranking individuals. The Old World patas monkeys and blue monkeys in the wild do not have dominance hierarchies probably because the dense foliage of the forest environment is a barrier to exerting control over others (Rowell, 1988). Patas monkeys in captivity do not have a clear unidirectional submission signal such as the silent bared-teeth display of rhesus macaques. This is consistent with the field observations of the species (Loy et al., 1993).

Dominance styles of different species have continued to attract discussion since the suggestion was made that there is a doubled-layered hierarchy; a fixed order marked by submission displays (for example, the bared-teeth signal) and a second level of more changeable actual relationships (de Waal, 1986b). Studies have shown that rhesus and Japanese macaques have strict hierarchies, whereas stumptailed and Tonkean macaques do not. Once again we are reminded of the sheer diversity in behaviour and environments.

Japanese macaques in captivity show low tolerance of the proximity of subordinates (Chaffin, Friedlen, & de Waal, 1995). There have been several recent studies of dominance in Japanese macaques focusing on their female dominance hierarchies (Nakamichi et al., 1995). In captivity studies of Japanese macaques, stable dominance relationships have been reported even among infants at the weaning period, taking their cues from their mothers' dominance-related actions (Chapais, 1996). Any generalizations from these studies need to be qualified to some extent by the acknowledgement that they are of one particular species under conditions of captivity. Even so, it is true that in some of the cercopithecinae (macaques, baboons, and vervet monkeys) there are matrineal dominance hierarchies in which females take on their mothers' rank.

MATRILINES

Typically, in captivity studies, macaques organize themselves by matrilines in which dominance status is determined by the status of the mother, grandmother, or great-grandmother. Within each matriline, the rank/dominance relationships between siblings of each family depend on birth order. Mothers are dominant to their daughters, but younger females usually rise in rank above their elder sisters. This pattern is known as Kawamura's principle after Kawamura (1958) who first called attention to it, in Japanese macaques, although it has not been found to hold in studies in the wild of this species (Hill & Okayasu, 1995).

So far as the dynamics of matriline hierarchies are concerned, young female Java macaques have to be social tacticians to attain dominance over adults from lower matrilines. The first (indirect) move towards asserting dominance is *appeal aggression*, when the juvenile, male or female, enlists support from elders. Here, head movements and glances at potential supporters occur simultaneously with menaces to the opponent (de Waal, 1977, 1987). At this early stage, the adult can still reply by threatening the juvenile; the latter retreats or gives the bared-teeth signal. As a juvenile reaches four years of age it is able to threaten an adult directly with a stare and open mouth. Gradually, the roles of sender and receiver of threats are reversed, the adult eventually avoids the juvenile and signals submission by teeth-baring. Even though the social structure determines the juvenile's starting position, the behavioural outcomes still have to be earned.

LONGEVITY AND DOMINANCE CHANGES

Some popular discussions of dominance orders have insisted on the beneficial effects of a hierarchy and suggest that it could be a model for human society. The essence of the argument is that the hierarchy should be stable, so that the social system is conflict-free.

Male dominance orders in primates are often not stable at all. In olive baboons, for example, males frequently migrate from their natal groups and others migrate in, so that there is a continual turnover. This means a rotation of dominance around the group and continual realignment of alliances and dominance orders. A further factor is that the longevity of species, up to 50 years for chimpanzees and 20–30 years for monkeys, also provides ample opportunities for dominance rotation. The dominance relationships of chimpanzees change as a result of challenges by males reaching adulthood; the deposed males eventually submit but can become allies later. In male chimpanzees, the prime period between 21 and 26 years of age is the particular time for challenging dominants. Altogether, over the life span, there are many such opportunities (Nishida, 1983). The dominance career of chimpanzee males follows an inverted U-shape curve; young males are lowest, prime males highest, and older males again lower (Bauer, 1979).

DOMINANCE AND GAINING A MATE

Alpha-ranked males in a number of baboon studies do not have the absolute priority in mating that earlier studies suggested, because of alliances among subordinate males and alliances of males with females and their infants. Mating patterns are the result of a complex contest in which the winners are not necessarily those of superior rank, this much is clear (Smuts, 1997).

Concerning priority in mating, the evidence on the advantages of dominance is considered uncertain (Smuts & Smuts, 1993). Dominant male chimpanzees rarely monopolize females and, if they attempt to do so, they are often unsuccessful. In macaques and baboons, even when alpha males guard females, they are not likely to be completely successful. Other males "steal" the females. Female choice is, no doubt, also involved and the concept of theft in this case is suspect because it assumes male control over females. In the discussion of the "klepto-copulator" stratagem in red deer (Clutton-Brock, Guinness, & Albon, 1982), the thieves were males who stole females while the alpha male was contesting a challenge from another. A thieving male might produce as many offspring as a dominant.

Similar sneaked copulations were reported in a study of captive Barbary macaques (Keuster & Paul, 1989). Although consortships did occur between adult males and females, it was more common for males to circulate among the females. Young males, although subordinate to adults, were able to mate with females at peak mating periods while the alpha male was momentarily distracted. This stratagem was possible due to the extreme rapidity of copulations. Opportunistic mating by younger males was frequent, usually when a dominant male was diverted by coalitions of rivals. In this situation, it was rare for females to hesitate in mating with young males. In many studies, opportunism of this kind has been reported and it has destroyed any absolute assumptions of preferential mating by dominants.

FEMALE COALITIONS AGAINST DOMINANTS

Adult male chimpanzees and baboons are always stronger than a single female, so females form protective coalitions. These coalitions occur in three situations: against strange males entering a group, defending a female attacked by a male, and for protection of infants (Smuts, 1987).

In captivity, the females of lower-ranking matrilines cannot escape harassment by dominants. In the wild, too, there are lesser life chances for lower-ranking female baboons: they give birth to fewer infants and raise fewer to maturity than dominants (Dunbar, 1984). Similar effects have been found both in captivity situations and in the wild, where reduced birth rates were found in lower-ranking females. In some New World species, such as marmosets and tamarins, this meant total suppression of fertility in subordinates, which then became helpers in rearing the offspring of the breeding female. There is convincing evidence for such effects on subordinate females' breeding rates, in New World monkeys and in macaques, baboons, and geladas (Abbott & George, 1991), although there is also evidence of the use of coalitions as a defence.

Dominant females harass lower-ranking females with threat signals, displacements, and attacks. Birth rates for low-ranking females that are partners in defensive coalitions are higher than for those that are not. The costs and benefits of forming coalitions provide a fascinating area for speculation (Dunbar, 1980, 1988). Costs are the physical risks involved in opposing a dominant; in the proximate situation, the actual benefits are protection from attacks and displacements. A low-ranking female has to choose a partner that will reciprocate consistently and will send this message to a dominant.

How is an observer to estimate the costs or benefits of coalitions? After aggression by dominants, no physical injury may be apparent even where lower-ranking females are threatened frequently and there are surely costs in terms of stress. It could be a choice between the costs of harassment against the costs of being an ally.

CONCLUSIONS

Physical size and strength are not the only influences contributing to dominance, nor to challenges to a dominant. Social skills are important because, in highly sociable primates, few fights are confined to two individuals. If individuals form coalitions they can choose the time for a fight to take full advantage of the presence of allies.

Dominance was seen in the 1960s as a universal principle in social organization, even though it was only clearly identified in ground-living primates. That view has changed. Researchers were unable to agree on how to measure dominance from the variety of indicators including access to sexual partners, food, drink, and space. Correlations between male dominance rank and mating were inconclusive over 40 years of research studies. Given the complex dynamics of coalitions and dominance-shifts, efforts to show general dominance effects seem to have undertaken a task that was just too difficult. Nevertheless, dominance is often a feature of the social activities of Old World monkeys and apes and therefore has a central place in the social psychology of primates. It is both a product and a cause of social structures that vary considerably across species, even between different species of macaques.

CHAPTER TWELVE

Conclusions: Primate Studies and Social Psychology

BREAKING NEW GROUND

I have discussed studies of the communication systems, signals, and social actions of monkeys and apes, which have provided much new information. Vocalizing had been thought to be expressive mainly of emotions and this has been shown to be a very limited view. In many studies, vocal sounds are now shown to achieve social goals, to influence or control others, and to send messages about events in the environment. The close-range vocalizing of monkeys often seems to be conversational in content: an exchange of messages (Dunbar, 1993). These sounds are graded, blending into each other, yet sending distinct signals to their hearers. In this characteristic, the sounds seem to qualify as precursors of human speech.

The social uses of vocal sounds are important for our fuller understanding of primate thinking. Eye-contact and gestures are also important signals; facial expressions, body movements, and touch have figured in the discussion, in previous chapters. Much of this is new knowledge which may change our concept of monkeys and apes. Researchers' accounts of negotiating, of deception, and other social tactics must surely have some influence, even on the moral stance to be taken towards the protection and conservation of nonhuman primates.

Studies of social behaviour suggest strongly that it is an error to place monkeys' and apes' social actions in a separate category of *animal behaviour* and therefore compartmentalized from human behaviour. There is a marked and traditional tendency in the discipline of psychology to continue to do this (see Domjan & Purdy, 1995, for comments on this tendency). On the contrary, it seems more realistic to regard the social behaviour of nonhuman primates as having special affinities with that of humans. The study of primate social actions should therefore be an integral part of a psychology that is valid within an evolutionary framework. Genetic, biochemical, and anatomical measures, while they reveal the differences between nonhuman and human primates, still confirm a view of primate distinctiveness, as contrasted to other mammals. It is, clearly, appropriate that communication of nonhuman primates be studied with the methods and concepts of social psychology.

Human rational choice models have had some influence on researchers' methodology although, from a social psychological position, they seem to focus narrowly on individual motives. Methodologically, we need observations of reciprocal, joint, interactive, and long-term social behaviour. Theoretically, we need explanations that always include the decisively formative and ever-present social context.

OPTIMALITY MODELS

A discussion by Hans Kummer (1978) will serve as an example of a rational choice model. He asks (p.687), referring to monkeys, "Why does an individual put up with the time-consuming frustrating and often harmful ado of group life rather than living . . . unhindered . . .?". Such a pre-social individual opts into social life only if gains outweigh costs. What have individuals to gain from entering into social relationships? For an economic-rational monkey, social relationships are investments which must bring benefits such as warding off rivals. The disadvantage of the model is its conceptual separation of an individual from the social environment, because individuals never are separate from their social environment. In this vein the model considers conflicts between an individual's interests and the disadvantages that social life brings. Obviously, this emphasis on a *pre-social individual* is somewhat defective if we are interested in the ever-present formative influences of joint and mutual relationships. Current work on primate social behaviour has not followed this approach too much, even if the model is valuable as a speculative effort.

This approach was similar to optimality models, familiar to zoologists, recognizing costs and benefits of behaviour and suggesting that traits evolve to be optimal. The best choice is the alternative in which advantages outweigh disadvantages translated into units of fitness; the preferred choice should maximize reproductive fitness (for example, the number of surviving offspring). The model can also be applied to short-term optimality—time, energy, factors that are thought to be related eventually to fitness. Natural selection will favour the optimal choice, the one in which benefits outweigh costs by the greatest margin. Foraging is a frequent subject for applying optimality theory because costs and benefits can both be assesed in energy units, net energy gain per unit of time (Goodenough, McGuire, & Wallace, 1993) and one decision can be considered at a time: where to eat, what to eat, how to search for it.

GAME THEORY AND SOCIAL EXCHANGE THEORY

Two-partner game theory models are clearly more appropriate for social behaviour and have been used effectively many times in zoology, primatology, and social psychology to suggest hypotheses for predicting patterns of social behaviour. So, Ronald Nöe's (1989, 1992) coalitionary model predicts *minimal resource* strategies for partners in baboon coalitions where the partners are both seeking the same resources. In this case, the preferred alliance is with the weakest partner available, as predicted by the minimal resource rule and explained by the advantage of the low competitive power of the ally. This was, in fact, borne out in the asymmetry of alliance outcomes that Nöe reported. One partner took the major share although the other partner must have derived some benefits from the alliance. According to Nöe, other examples of rule-following, as predicted by game theory, might be found in the signalling episodes that seemed to reflect bargaining, acceptances, and refusals between prospective partners. Partners might also be signalling their discontent with the existing balance of outcomes, by means of threats to discontinue the alliance.

Proposing game theory models, comparing them to actual patterns of coalitionary behaviour, should allow us to explore the proximate level of communication, throwing more light on cognitive abilities. Is there rule-following in the social tactics of rivals and allies? Coalitionary actions seemed to follow a simple trial-and-error learning rule, but Nöe's descriptions of bargaining sequences in which there was an exchange of gestural, tactile, and vocal signals suggested that the baboons were informing each other of their intentions as future partners. We can be confident that studies of this kind, showing flexible stratagems that are matched to the actions of others, are highly relevant for our future understanding of baboons' social intelligence.

Potentially, social exchange theory is another relevant approach to understanding the social actions of nonhuman primates, although often it is an implicit rather than an overtly stated theoretical framework. This theory has both a behavioural and a social psychological emphasis in the sense that explanations are tested by observing the reciprocal actions of partners. It also has the merit of drawing attention to cognitive capacities of recognition and memory which are necessary for maintaining balanced exchange relationships over periods of time. Social exchange ideas as they currently appear in social psychology (Foa et al., 1993, Hinde, 1997) have a strong cognitive emphasis in so far as the perceptions of the partners to an exchange, their evaluations of each other and of the items exchanged, are often considered decisive factors. In the past, the inheritance of behaviourist ideas had been limiting to the development of the theory, notably after influential inputs from reinforcement and operant learning theories in the 1960s and 1970s. This influence tended to reduce the domain of the theory to individual reinforcement contingencies, but current discussions of social exchange ideas give more central importance to understanding the reciprocal relationship itself. This development is perfectly well suited to the contemporary logic in theorizing on such topics as primate grooming relationships and their longer-term joint consequences: why would individuals undertake grooming partnerships "if not to foster further beneficial exchanges?" (de Waal & Aureli, 1996, p.88).

SCIENTISM AND SOCIAL PSYCHOLOGY

Some studies of primate social behaviour, also of the 1960s and 1970s, took a diversionary path into a scientism that raised techniques to an exaggerated degree of importance, to the detriment of understanding the social actions themselves and to the exclusion of the meaningful context of messages. Stuart Altmann's (1965) work at that earlier period seemed to have emphasized quantification at the expense of understanding whatever meanings were to be decoded from social interactions. This was a relatively unproductive path into which others followed (see, for example, Maxim, 1976).

Peter Slater (1973) pointed to the difficulties in making sense of Altmann's study of rhesus monkeys. The analysis did not distinguish between effects from an individual's previous actions and effects from the actions of the partner. This confusion would be a severe problem for a social psychology of rhesus monkeys. Altmann's study focused on a statistical analysis of actions irrespective of which monkey had made them. As a result, "two consecutive actions (might) be by the same or by different individuals" (Slater, 1973, p.148). Slater gently understated the problem:

> confounding of these two effects was perhaps inevitable in a field study where several animals might be in view and each could show any one of 120 behaviour patterns, but it means that his (Altmann's) results cannot be interpreted easily.

A search for meaning, uncertain as this goal often is, must be the main influence on decisions about the actions and events we consider worthy of study and how we study them. In the social psychology of monkeys and apes we study social events, action by action. Most of the empirical basis for discussion consists of the flexible tactics of social relationships. At the same time, there are some problems in finding information that is comprehensive on social actions. For example, negotiation is a social activity that has recently been mentioned by several researchers. At the moment, it is difficult to establish how exactly the negotiation takes place, and descriptions have been relatively brief. There is an encouraging convergence in researchers' reports but we need to know more about the negotiation process itself. All this promises a rich area for social psychological research in future studies.

A social psychological approach has required that we view signals as operating together in sequences. Many primate studies have tended to concentrate on one channel (for instance, vocal sounds) without considering signals as they are used in combination. “How does social interaction happen?” has been our guiding question and the answers have sometimes been elusive. Questions that have been raised about the abilities of monkeys and apes often have only tentative answers. Some writers have been pessimistic or perhaps impatient. As Hans Furth (1992, p.156) comments, in nonhuman primate studies “there is not a single exact answer concerning the precise cognitive mechanisms involved in any particular know-how.” Furth was referring here both to observational and to experimental studies; in each case results are inconclusive, of course. Yet this is a commonplace situation wherever scientific research on cognition and social behaviour is undertaken. Andrew Whiten (1996b, p.4), discussing the disagreements over the possibility of a cognitive gap between monkeys' and apes' abilities, concludes that

> this particular branch of evolutionary psychology is at an intellectually exciting stage where many fine minds are trying to discuss the larger patterns in what is now a very rich web of information. So long as those reaching disparate conclusions are prepared to converse . . . we can continue to look ahead to convergence on reliable scientific knowledge.

Recent studies have provided continuous descriptions of social interactions with full accounts of signals, vocal and gestural. But there seems to be a methodological problem over whether social interaction descriptions can be incorporated into designs for research programmes. It is only very recently that a variety of signals in primate communication (sounds, gestures, eye-contact and so on) have been included in research reports. These developments, as discussed in previous chapters, have been especially exciting because they have crossed new ground, They give insights into areas of social behaviour that have been almost completely unknown, insights that may have an important influence, eventually, on our views about the nature of other primate species.

AN EVOLUTIONARILY VALID PSYCHOLOGY

There are two comments that I want to make on the coverage of new topics in primate communication. First, although this knowledge is accumulating very rapidly, the extent of the advance in primate studies is not generally known to professionals in the disciplines of sociology and psychology. Second, although many investigators have opened doors and avenues in the methods and concepts used to study social behaviour, the social psychological perspective is little understood in primatology. For instance, discussion of the intriguing work on chimpanzees' recognition of their body-image in a mirror will be more adequate if the self and self-image are recognized as social constructions. Perhaps the essence of the primate self is to be discovered in the study of social relationships, in so far as the self is formed and expressed through them. As more of these relationships are revealed, so we will know more of the complexity and limits of nonhuman primates.

Much of the research material concerning social activities discussed in preceding chapters, has implicated *social intelligence*—in contrast to the *technical intelligence* that manifests itself in the manipulation of objects (Byrne, 1995, 1997). Social intelligence is expressed in communication and I have given examples of the complexity of social tactics. The use of intentional movements and signals to achieve social objectives has been especially emphasized. An individual tried to gain a social objective: if one action did not succeed then another was tried out; a variety of actions might be used, all with the same objective in mind. We should understand not only that individual *A*'s action influenced the actions of the other individual but that *A* knew how to produce that effect on *B*'s behaviour, how to influence *B*. *A* knew how the action would be taken, that *B* would react in a certain way.

In social interactions, the use of sounds, hand gestures, facial expressions, and body movements repeatedly affirm the affinities between monkeys, apes, and humans. One reason why these signals are important in the evolutionary process is because they constitute the social behaviour from which language must have emerged. Signals are not only reflexive (in the sense of expressing emotions) but have symbolic content, sending messages to the hearer about things, events, intentions. Speculations about the gestural contribution to the development of spoken language have emphasized the importance of hand gestures, as ancestral hominids became fully bipedal and hands became free both for tool-use and for more gestural communication. Arguments over this possibility focus on the problem of how an evolutionary cross-over could have occurred from the assignment of meanings by hand gestures to, eventually, speech and syntax (Armstrong, Stokoe, & Wilcox, 1995; Bickerton 1990). Either gestures or sounds may have provided the beginnings for structures of meaning that later became language: these are contentious issues that I have avoided here. I wanted to show, from studies of social interaction, how signals are used and combined in situational context and what effects they produce. There are a sufficient number of questions arising in this area to give ample scope for new research efforts grounded on the shared concerns of primatology and social psychology, and contributing to theory and explanation in both disciplines.

APPENDIX ONE

Primate Species and Their Social Structures: Lesser and Great Apes, Old World Monkeys, and New World Monkeys

This appendix gives a broad summary of the social organization of the primate species that are discussed in the book and presents them in the systematic form of families and subfamilies of species. When I refer to a species I also give their Latin names, following the convention used by primatologists. In the previous chapters, Latin names were given at the first mention of the species.

Old World monkeys, apes, and humans share a common characteristic in that they have long, socially complex lives (Zihlman, 1997) but even in this outline-summary of social structures, the enormous diversity of primate species will soon become apparent.

Nonhuman primates are classified into three broad types or genera: the Lesser and Great apes, Old World monkeys, and New World monkeys.

LESSER AND GREAT APES

Apes are larger in size than monkeys and differ from many monkey species in that they have no tail [but Barbary macaques (*Macaca sylvanus*) and Japanese monkeys (*Macaca fuscata*), species that live at higher altitudes, have no tail and other ground-living species have shortened tails]. Apes are closest to humans in intelligence, blood composition, and anatomy. The lesser apes, the gibbons (*Hylobates moloch*) and siamangs (*Hylobates syndactylus*) in the rainforests of South East Asia, are arboreal and their basic social unit is the male–female pair (Zihlman, 1997). They are the most numerous of the apes but perhaps the most difficult to study in the wild. Even their species classification has not been conclusively agreed. Some primatologists think that there are five, but most that there are nine species of gibbon in all, differing in colour shade and markings and in their calls. Gibbons and siamangs have enduring pair relationships and their group size can reach a maximum of five, consisting of the breeding pair and developing offspring, but when this number is reached the oldest maturing offspring leaves.

The great apes include two species of chimpanzee; the common chimpanzee (*Pan troglodytes*) and the bonobo or pygmy chimpanzee (*Pan paniscus*). There are three species of gorilla (*Gorilla gorilla*). All of these are largely ground-living (terrestrial). The remaining great ape species, the orangutan (*Pongo pygmaeus*) of Borneo and Sumatra, is predominantly tree-living (arboreal).

The social structures of the great apes are extremely varied. Adult orangutans are virtually solitary except for mothers and dependent youngsters. Their few social contacts consist only of brief associations between males and females for mating and occasional meetings between females and juveniles. Adolescents are relatively sociable, associating together for days or weeks at a time. Orangutans are acknowleged to be the least gregarious of the apes and studies agree that their social units are (1) solitary adult males, (2) adult females with offspring, and (3) solitary subadults. Occasional larger groupings occur at food sources (Rodman & Mitani, 1987). Mitani (1985) observed associations between subadult males and adult females over periods lasting up to 28 days, and adult males and females formed consortships for as long as 5 days. As Mitani, Grether, Rodman, & Priatna (1996) noted, associations among orangutans are ephemeral in comparison with most diurnal primates. Orangutans have a relatively simple social system although it is possible that a more varied social network may emerge from current studies.

Gorillas live in small groups, with a variable number of resident males, from one to many. A group may comprise about a dozen members including a silverback adult male, other males, females, juveniles, and infants. Gorillas have also been observed in all-male groups; one group of six males was followed and observed by Juichi Yamagiwa (1987) for a period of about three years.

It is difficult to make summary statements about social structures without some inaccuracy. It is true that "gorillas live in relatively stable bisexual groups" (Stewart & Harcourt, 1987, p.156) yet solitary animals, usually adult males, make up about 10% of most populations. Stewart and Harcourt pointed out that the group size in West Africa (median size 5, maximum observed 12) is generally smaller than in Eastern populations (median size 9, maximum 37). In areas for which data exist they noted that 60% of groups have only one adult male. As I have mentioned, there are also all-male groups.

Chimpanzees have a social structure that is flexible and complex. Individual females and infants and small bands of males associate together in communities of up to 80 members; the average size is just over 50 (Dunbar, 1993) Bonobos have a similar social structure but their characteristic mixed group consists of males and females of all ages. Bonobos are a rare species living in one region of Zaire; they are of slighter build than common chimpanzees, hence the name pygmy chimpanzee. They have only been studied closely in the wild since the mid-1970s. Female bonobos are more sociable than either female gorillas or female chimpanzees, spending less time on their own. This may be because they are sexually active over much longer periods than chimpanzee females. In fact, sexual activities of various kinds between females and males, between females (genital rubbing), and between males are more frequent in bonobos. This has been noted in studies in the wild (see Susman, 1984; White, 1996) and confirmed in captivity studies (de Waal, 1988). The strong bonds among females are expressed frequently in female coalitions in this species in which male coalitions are rare (Kano, 1992).

Knowledge of Social Behaviour

26 out of 71 Old World monkey species have been classified as "uncertain" as far as systematic knowledge about their social lives is concerned (Smuts et al., 1987). As can be imagined, the uncertainties about arboreal New World monkeys are relatively greater: concerning their social organization, 24 out of 47 were marked as "in doubt" or "unknown".

Apes and Old World monkeys have figured centrally in the discussion of the social psychology of nonhuman primates. Among the apes and Old World monkeys, species such as chimpanzees, gorillas, baboons, and macaques have been studied most. Even in apes, however, it is extremely difficult to collect information on the social interactions of the rapidly moving and elusive arboreal gibbons and orangutans. Classic studies exist of the gibbon by Ray Carpenter (1940) and of the orangutan by Biruté Galdikas (1978, 1984, 1988, 1992), but detailed accounts of the species' social behaviour are scarce.

Core Groups: Female and Male Migration

At adolescence, gorilla, chimpanzee, and bonobo females often transfer out of the groups in which they were born. Jane Goodall (1986) referred to three kinds of female transfer among chimpanzees—permanent transfers or *immigrants*; temporary transfers or *visitors* (females who visit neighbouring communities for short periods and then return); and *peripheral* females who move repeatedly between one community and other.

This transfer of females in gorillas and chimpanzees contrasts with the pattern in many monkey species where it is the young males who migrate, often to join so-called bachelor bands, and the females are phylopatric, remaining in their natal groups. Actually, both young male and female gorillas migrate from their natal units. In colobus (*Colobus* species) and howler monkeys (*Alouatta* species) too, both sexes migrate. In orangutans, males move to new areas to meet females. As orangutans are mainly solitary there are no groups to transfer from or into. Galdikas (1978) who studied orangutans in Borneo for five years, argued that female selection of mates was more influential in orangutan mating behaviour than it was in gorillas and chimpanzees. She mentioned as the only common attribute of social organization in the great apes (gorillas, chimpanzees, and orangutans), the general lack of social bonds between adult females (in marked contrast to many monkey species). It is characteristic of the dynamic state of knowledge about nonhuman primates that her statement has to be revised to take into account more recent information. Bonobo females do have strong social bonds—they share activities much more than females of the other great apes. At the core of their groups, in fact, are strongly bonded females. (Kano, 1980, 1989). Although chimpanzees and bonobos both move in and out of flexible, temporary associations, the mixed male/female social group is still more typical of bonobos (White, 1992a, 1992b, 1996).

Mutual grooming could be taken as an index of social bonds between individuals. Grooming involves parting and searching through the hair with the hands and sometimes with the mouth. Hygienic in function, it is obviously also an important social activity. Chimpanzee females participate in mutual grooming very little, compared to males (Simpson, 1973). After three years of field studies at Kasoge in Tanzania, it was reported that 46% of all grooming occurred between chimpanzee males and 39% between females and males, but only 10% between females (Nishida, 1979).

Galdikas had observed that female chimpanzees and female orangutans led a mostly solitary existence with the companionship only of their offspring, but even the most solitary chimpanzee female was still much less solitary than a female orangutan. However, the greatest differences between orangutan and chimpanzee sociability is seen in the adult males. Chimpanzee males are more sociable than females, adult male orangutans are more solitary than females. A unique social characteristic of orangutans is the absence of any positive relationships between adult males. In fact, they are extremely intolerant of each other.

OLD WORLD MONKEYS

The Old World monkeys of Africa and Asia can be divided into two subgroupings. First, the *cercopithecine* monkeys that are mostly tree-living in Africa; these are the mangabeys and guenons that are mainly restricted to forests. Also in this subgrouping are the partly ground-living (terrestrial) macaques (*Macaca* species) of Asia, Africa, and Japan and, finally, the entirely terrestrial baboons (*Papio* species) of Africa.

The second subgrouping are the *colobines* that are largely arboreal. These are mainly found in Asia (langurs, *Presbytis* species), with a few colobus species in Africa. Some primatologists distinguish the two subgroups by referring to the cercopithecines as the "guenon" subgroup and to the colobines as the "leaf-monkey" subgroup.

The *cercopithecines* consist of the monkeys that are best known. They have a very wide geographical distribution; from Gibraltar and North Africa (Barbary macaques *Macaca sylvanus)* across Central and Southern Africa (five species of baboons *Papio species*), across Asia (especially rhesus monkeys *Macaca mulatta*, of India and Pakistan).

Among the cercopithecines are the vervet or green monkeys (*Cercopithecus aethiops*), the most widespread species. They live in African sparsely wooded savanna environments and are classified as semi-terrestrial. Patas monkeys (*Erythrocebus patas*) are also ground-living in the African grasslands. Many of the cercopithecine monkeys are tree-living and less available to close study in the wild. For that reason they do not figure in detailed studies of social activities. The most closely studied monkeys—baboons, rhesus, Japanese, and Barbary macaques, vervet monkeys and so on, all ground-living species—take centre stage. Japanese macaques are typical of macaque species in that they live in multi-male, multi-female groups characterized by females that remain in their natal groups and males that migrate on puberty to other groups.

There is a great variety in the social organization of cercopithecines. Some species, such as hamadryas baboons live in groups that consist of one adult male with several adult females, adolescents, juveniles, and infants.

Ronald Hall and Irven DeVore (1965) described the typical hamadryas social structure which they explained by their "protection theory" (see Rhine, 1975; Rhine & Westlund, 1981). In a troop on the march, dominant males guard females, with the infants and juveniles in the centre. The subordinate male followers are at the front and rear of the troop like outriders. When the troop is resting, the females are still at the centre and the males guard the perimeter. In defence against predators, the dominants move to the front and lead the other males into attack, while females and young go to the rear.

Hamadryas social organization involves two types of social bonds; a bond between the adult male and his female companions, and a bond between adult males. "Bond", in this context, refers to a relationship based on (a) mutual attraction, and (b) the specific exclusion of others. In hamadryas baboons, the male–female bond is aggressively enforced by males (Kummer, 1968).

Other species of baboons do not have the same distinctive one-male system as hamadryas baboons. Relatively little has been reported on the Guinea baboon but for the other three baboon species and for geladas (*Theropithecus gelada*) a very large amount of field and captivity data has been reported. Three of the baboon species, the olive (*Papio cynocephalus anubis)*, the yellow (*Papio cynocephalus cynocephalus*), and chacma baboons (*Papio cynocephalus ursinus*) are known as *savanna baboons*. Their social units are considerably more flexible than the hamadryas one-male units. These three species extend from Northern to Southern Africa and live in large male/female groups of 20–200 with a breeding system in which both males and females have several or many partners.

Geladas (*Theropithecus gelada*), a species closely related to baboons and sometimes called gelada baboons, also have a social structure that is centred on the one-male unit but it is much less rigid than the hamadryas unit. The gelada one-male unit frequently disperses, unlike the hamadryas unit that stays together at all times. Vernon Reynolds (1984) discussed these differences, pointing out that the gelada male and his females form a loose and sometimes scattered unit in the troop. Reynolds saw this as another characteristic shared with early (hominid) forms of human groups, which possibly developed into a pattern of males leaving their females to go hunting. Geladas live in one-male, multi-female groups but the females are not herded together by the male, as happens with the hamadryas. In olive and yellow baboons, both sexes live together in multi-male groups of several adult males and many females, adolescents, juveniles, and infants.

Among the savanna baboons, females form the core of the social system. The fact that female savanna baboons try to avoid other females during foraging suggests that there are competitive relationships between females, just as there are between the males.

One feature common to many cercopithecine species is adolescent or subadult males migrating from their natal group, sometimes migrating again at a later point in the life-cycle (Sapolsky, 1996).

The other subfamily of Old World monkeys, the colobines, live mostly in Asia, but there are nine species in equatorial Africa. Males and females tend to be less sexually dimorphic (less different in body size) than terrestrial monkeys. The exception to this is the proboscis monkey *Nasalis larvatus*. Males in this species weigh more than twice as much as females.

Colobines may form large multi-male groups of up to 40–50 individuals and larger groups (up to 350) have been reported (Cords, 1993b; Snowdon, 1993) but, again, it is impossible to generalize across all the subfamily. Some species form one-male, multi-female groups or male/female pairs. As far as any typical group composition is concerned, research has shown that, depending on environment, there is great variation even within a given species.

Some species have a flexible social organization which makes it difficult to apply categories such as *one-male units*. Patas monkeys, for example live in their one-male, multi-female groups only *outside* the short mating season. For the duration of this one-month mating season, small bachelor groups join females for mating, after which the patas social unit resumes its one-male composition for the rest of the year. As will have become clear by now, it would be unsafe to generalize across primate species about group composition or size.

NEW WORLD MONKEYS

The New World monkeys of Central and South America consist of two sub-groupings: the *callithrichids* (marmosets and tamarins) and the larger size *cebids* (capuchin and howler monkeys). The callithrichids are very small, have claws on all digits except the big toe, and are entirely arboreal, and, for this reason, are also hard to observe in the wild. Captivity studies usually focus on male/female pairs. They, typically, give birth to twins and breeding takes place every two years; an adult pair may have several twins to look after before the first set are mature enough to leave. Thus, a family group may have up to eight individuals. Field studies recently have shown that marmosets and tamarins are not necessarily monogamous in the wild; polyandrous, polygamous, and monogamous mating patterns have all been observed (Digby, 1995).

Cebids, the "true monkeys" of the New World, have a varied social organization. Some, like titi monkeys (*Callicebus* species) live in monogamous pairs. Others live in groups of several adult males and females, although individuals forage independently. The larger cebids tend to be polygynous. Capuchin monkeys (*Cebus* species) live in multi-male, multi-female groups of up to 35 individuals. The large howler monkeys (*Alouatta* species) live in multi-male, multi-female groups of 10–20. The small squirrel monkeys (*Saimiri* species) band together in groups of 20–40 members. Sheer numbers enable them to hold their own when competing for food with larger monkeys.

TERRESTRIAL AND ARBOREAL PATTERNS

Four distinct basic structures of social organization that include both terrestrial and arboreal species have been tentatively suggested (Dunbar, 1988). Table A1 is derived from this four-fold classification.

In suggesting this broad classification, Robin Dunbar mentioned the difficulty of fitting the species into it. The one-male and several-female groups of colobus monkeys (*Colobus* species), hamadryas baboons (*Papio hamadryas*), and geladas (*Theropithecus gelada*) also belong to larger communities of 60–100 individuals. They can be classified as one-male *and* as multi-male, multi-female

TABLE A1
Social Structures

Type	*Example*
1. Semi-solitary	Orangutans (a)
2. Male/female pairs	Marmosets (a) gibbons (a)
3. One-male/several-female (harem) groups	Colobus monkeys (a), (t) gorillas (t) hamadryas baboons (t) gelada baboons (t)
4. Multi-male, multi-female groups	Savanna baboons (t) macaques (t), (a) chimpanzees (t), (a)

Adapted from Dunbar, 1988.
(a) = arboreal (t) = terrestrial

groups. This has been called a *multi-level* social structure. The one-male groups can merge into larger groups or disperse again into small units, depending on the needs and activities of individuals. In many species there are variations in social structure which depend on the ecological factors of different locations. It is impossible to summarize all these variations here and information is still lacking for some species. Chimpanzees fail to fit into the classification easily. They are both terrestrial and arboreal; they may form male/female consortships (as do savanna baboons *Papio cynocephalus* species) for limited periods, while females-and-infants may be solitary much of the time.

NOTE ON THE FISSION–FUSION GROUP

The influence of ecological factors on the social organization of primates has been an important theme for many studies. Availability of food sources has contributed to variations in social patterns (Chapman, Wrangham, & Chapman, 1995). Explaining why social organization takes a particular form in nonhuman primates is usually a matter of speculating about evolutionary pressures, such as defence against predators or competition for food. If food is available in small dispersed quantities as it is in the African savanna, then small feeding units will be most suited to this dispersion. A convincing rationale for changes in social organization can be found in these two kinds of pressure. For example, the hamadryas multilevel group consists of small units, when foraging, that converge into large troops at resting places, forming a defence against ground-living predators.

Hans Kummer (1968) was the first to use the concept of the fission–fusion group. He identified within the hamadryas troop not only the one-male unit, but also travelling parties of two or more one-male units that stayed together over periods of days, and bands that consisted of several one-male units joining together in opposition to bands from other troops. The social organization of hamadryas baboons, therefore, illustrates a multilevel fission–fusion structure, superficially similar (in this characteristic only) to the social organization of chimpanzees.

However, chimpanzees and bonobos have a more far-reaching form of fission–fusion than hamadryas baboons, because each individual moves in and out of relationships and groups of varying size, from dyads to large parties, and does so in a much more flexible fashion, apparently from individual choice, rather than by the daily pattern of hamadryas reunions at resting places. For these reasons the chimpanzee and bonobo type of fission–fusion is a rare social pattern, shared only by the New World spider monkeys, *Brachyteles* and *Ateles* species (Hohmann & Fruth, 1995).

In chimpanzees, fusion and fission are taken to the limits of flexibility and the social activities of a chimpanzee are far more variable than those of most other primates (Goodall, 1986).

MOTHER–INFANT RELATIONSHIPS

Is there an identifiable social nucleus in primates despite the evident differences in social structures? Certainly, the mother–infant relationship as the basis of all primate societies has been emphasized by many studies. Paul Simonds (1974) referred to three major influences on social patterns in primate groups: (1) the sexual relationship, (2) the mother–infant relationship which he regarded as the primary social bond, and (3) the social bonds between adult males, in gorillas, chimpanzees, and baboon species in which females rather than males, migrate out of their natal groups. To these can be added, (4) the long-term alliances between females in *female-bonded* species which comprise the vast majority of primates (Wrangham, 1980). In these species the males migrate, and female-led matrilines or families constitute the stable social group.

In monkey species, the mother–infant relationship often has a lifetime influence. A mother will support her offspring in contests with others so that the relative dominance of the mother perpetuates itself in her offspring. Daughters of high-ranking females tend to maintain the maternal rank into adulthood. A colony of rhesus monkeys (*Macaca mulatta*) introduced into the small island of Cayo Santiago in the Caribbean has been studied over decades and the stability and transmission of maternal status has been amply confirmed (see also de Waal, 1996b). Offspring continue to associate with the mother and siblings over the total lifespan. In the large troops that are characteristic of chimpanzees, the primary stable subgroup is also the mother–offspring unit but, young chimpanzees, as they reach maturity, spend more time with other groups, and males, in particular, associate with other males (Morbeck et al., 1997).

FACE-TO-FACE CONTACTS

In a social psychological approach, the focus of study is on face-to-face contacts at a level of close detail. For primatologists, social structures usually refer to sexual and infant-rearing relationships; male–female composition of groups, dominance relationships, grooming relationships, and location and movement of groups in a socio-ecological environment. As we have seen, many studies in these areas of social behaviour are also potential areas for developing a social psychological focus. One of the earliest attempts was Walter Angst's (1975) applications of the hypothesis that "the more two individuals interact, the more they will like each other". This was the well known interaction–liking hypothesis, argued over by researchers in sociology and social psychology in the 1950s and 1960s. In spite of its satisfying plausibility, the interaction–liking hypothesis was always controversial due to the long list of exceptions to it. Angst felt that it was useful for understanding the social actions of long-tailed macaques (*Macaca fascicularis*) that he studied in Java, Bali, and Basel Zoo. Zoologists have found the hypothesis attractive in several contexts (Boorman & Levitt, 1980) and I mention it here as a final example of interdisciplinary enrichment.

APPENDIX TWO

The Order: Primates

Comparisons of the social actions of monkeys and apes with those of humans lead us to consider the biological relatedness of nonhumans and humans. Considerations of the relatedness of monkeys, apes, and humans usually start with the taxonomy of the naturalist Carl von Linné (Carolus Linnaeus) who, in his *Systema Naturae* (1758), classified members of the Order: *Primates* by distinctive anatomical features (four incisors, two collarbones, an opposable first digit in hands or feet, and so on). On this basis Linnaeus divided Primates into four genera with the Latin names: *Homo*, (humans, orangutans), *Simia* (monkeys and apes, including chimpanzees), *Lemur* (lemurs and other "lower" forms), and *Vespertilio* (bats).

The Linnean system was modified by the anatomist St. George Mivart (1873), who proposed the removal of bats from the Order and reorganizing the remaining members into two suborders: *Prosimii* (pre-monkeys) for lemurs and *Anthropoidia* for monkeys, apes, and humans. Further modifications occurred later (Napier & Napier, 1967; Conroy, 1990). These have emphasized many other evolutionary characteristics of primates such as the enlargement of the brain, atrophy of the olfactory sense, prolongation of the postnatal growth period, and specialization of the hands or feet for grasping. Numerous other factors have been introduced into the classification. These factors are based on relatively new evidence on the structure of the chromosomes carrying the genetic material (deoxyribonucleic acid or DNA), the structure of the DNA itself and the proteins that it generates (Martin, 1986, 1990).

The similarities shared with humans have their most obvious visible expression in the head, eyes, and hands of chimpanzees, gorillas, orangutans, and macaques. The facial expressions and gestures of monkeys and apes send human observers a clear message of affinity. There are many such hints of similarity in the social behaviour of primates; comparisons between monkey, ape, and human are inevitable and follow a very respectable tradition.

A number of interesting parallels between human and nonhuman facial expressions and gestures were illustrated by Eibl-Ebesfeldt (1972). These were criticized because they could lead to a facile discussion of superficial resemblances (Martin, 1974). I agree that apparent similarities should be treated cautiously. Their significance has to be sought in the context of overall patterns of behaviour. If gestures, for example, occur as part of stratagems suggesting parallels between primates in social intelligence, then the similarity would be worth emphasizing.

GENETIC RELATEDNESS

When compared to the genetic distance of humans from other species, the human–chimpanzee difference is very small, corresponding to the genetic distance between sibling species of *Drosophila* or mammals (King & Wilson, 1975). Molecular biologists who have considered the evidence that humans and chimpanzees are very close in DNA also point to the many differences in anatomy, body organs, and behaviour that seem to belie the almost identical biochemical measures. The genetic closeness of humans and chimpanzees has to be reconciled with the actual differences in brain size, jaws, pelvis, feet, relative lengths of limbs and digits, and, of course, in behaviour. Nearly every chimpanzee bone is distinguishable in shape from human equivalents (King & Wilson, 1975). There are also the obvious differences in posture and gait and other differences in body organs: in vocal apparatus, palate, and digestive tract.

Pending further research, the answer to the apparent inconsistency between the genetic, biochemical evidence and the actual physical and behavioural divergence might be found in the gene regulatory mechanisms controlling the timing of development. *Control genes* regulating the development of the size and shape of the body's organs might provide the answers (see also Gould, 1977; Passingham, 1982).

BRAIN SIZE AND DIFFERENCES

Despite the close genetic relationship between humans and other primates, especially the great apes, the gap between humans and nonhumans in the size and structural development of the brain is evident. The human brain is much larger than the brains of nonhuman primates of the same weight. If the comparison is made between primate brains (human and nonhuman) and other mammals, however, the primate brain is seen to be distinctive. The human brain is now revealed to be a larger version of the chimpanzee brain, save for the areas of the neo-cortex associated especially with the specialization of language (Passingham, 1982).

It should be noted that humans are born with brains that are relatively small compared to their adult size. In newborn macaques the brain reaches 60% of its adult weight and, in chimpanzees, 46%. In newborn human babies the brain is 25% of adult size. Only in humans does the brain continue to grow, after birth, at the same rate as it did prenatally; this growth continues for about two years (Passingham, 1982).

Estimates of evolutionary development indicate that the ancestral lines of apes and humans diverged 4 to 7 million years ago, and that the monkey and human lines diverged 20 to 50 million years earlier. Fossil evidence of human-like primates dates from an estimated 3.5 million years ago, thus confirming the estimates from biochemical analysis of blood proteins (Passingham, 1982; Sarich, 1984; Sarich & Cronin, 1976).

References

Abegg, C., Thierry, B., & Kaumanns, W. (1996). Reconciliation in three groups of lion-tailed macaques. *International Journal of Primatology, 17*, 803–816.

Abegglen, J.-J. (1984). *On socialization in Hamadryas baboons. A field study*. Lewisburg: Bucknell University Press.

Abbott, D., & George, L.M. (1991). Reproductive consequences of changing social states in female common marmosets. In H.O. Box (Ed.), *Primate responses to environmental change*, (pp.295–309). London: Chapman & Hall.

Allen, C. (1995). Intentionality: Natural and artificial. In H.L. Roitblat, & J.-A. Meyer (Eds.), *Comparative appproaches to cognitive science*, (pp.93–110). Cambridge, MA: MIT Press.

Altmann, S.A. (1962). A field study of the sociobiology of rhesus monkeys (*Macaca mulatta*). *Annals of New York Academy of Science, 102*, 338–435.

Altmann, S.A. (1965). Sociobiology of rhesus monkeys II. Stochastics of social communication. *Journal of Theoretical Biology, 8*, 490–522.

Altmann, S.A. (1968). Social behaviour of anthropoid primates: Analysis of recent concepts. In E.L. Bliss (Ed.), *Roots of behaviour. Genetics, instinct and socialization in animal behaviour*, (pp.277–286). New York: Hafner.

Angst, W. (1975). Basic data and concepts on the social organization of *Macaca fascicularis*. In L.A. Rosenblum (Ed.), *Primate behavior*, Vol. 4, (pp.325–386). New York: Academic Press.

Antinucci, F. (1989). Systematic comparison of early sensorimotor development. In F. Antinucci (Ed.), *Cognitive structures and development in nonhuman primates*. Hillsdale, NJ: Erlbaum.

Arcadi, A.C. (1996). Phrase structure of wild chimpanzee pant hoots: Patterns of production and interpopulation variability. *American Journal of Primatology, 39*, 159–178.

Argyle, M. (1988). *The psychology of interpersonal behaviour* (4th Edn). Harmondsworth, UK: Penguin Books.

Armstrong, D.F., Stokoe, W.C., & Wilcox, S.E. (1995). *Gesture and the nature of language*. Cambridge: Cambridge University Press.

Asquith, P.J. (1981). *Some aspects of anthropomorphism in the philosophy underlying Western and Japanese studies of primate behaviour*. D. Phil Thesis. University of Oxford.

Asquith, P.J. (1984). The inevitability and utility of anthropomorphism in description of primate behavior. In R. Harré, & V. Reynolds (Eds.), *The meaning of primate signals*, (pp.138–176). Cambridge: Cambridge University Press.

Astington, J.W., & Gopnik, A. (1991). Developing understanding of desire and intention. In A. Whiten (Ed.), *Natural theories of mind. Evolution, development and simulation of everyday mindreading*, (pp.39–50). Oxford: Basil Blackwell.

Aureli, F. (1992). Postconflict behaviour among wild long-tailed macaques (*Macaca fascicularis*). *Behavioural Ecology and Sociobiology*, *31*, 329–337.

Aureli, F. (1994). Post-conflict social interactions among Barbary macaques (*Macaca sylvanus*). *International Journal of Primatology*, *15*, 471–485.

Aureli, F., & van Schaik, C.P. (1991). Post-conflict behavior in long-tailed macaques: II. Coping with uncertainty. *Ethology*, *89*, 101–114.

Badrian, A., & Badrian, N. (1984). Social organization of *Pan paniscus* in the Lomako Forest, Zaire. In R.L. Susman (Ed.), *The pygmy chimpanzee. Evolutionary biology and behavior*, (pp.325–346). New York: Plenum Press.

Balda, R.P., & Kamil, A.C. (1988). The spatial memory of Clark's nutcrackers (*Nucifraga columbiana*) in an analogue of the radial arm maze. *Animal Learning and Behaviour*, *16*, 116–122.

Baldwin, J.M. (1902). *Social and ethical interpretations in mental life*. New York: Macmillan.

Baron, R.A., & Byrne, D. (1991). *Social psychology. Understanding human interaction* (6th Edn). Boston: Allyn & Bacon.

Bauer, H.R. (1979). Agonistic and grooming behaviour in the reunion context of Gombe Stream chimpanzees. In D.A. Hamburg, & E.R. McCown (Eds.), *The great apes*, (pp.395–403). Menlo Park, CA: Benjamin/Cummings.

Bauers, K.A. (1993). A functional analysis of staccato grunt vocalizations in the stumptailed macaque (*Macaca arctoides*). *Ethology*, *94*, 147–161.

Bauers, K.A., & de Waal (1995). "Coo" vocalizations in stumptailed macaques: A controlled functional analysis. *Behaviour*, *119*, 143–160.

Beach, F.A. (1976). Sexual attractivity, proreceptivity and receptivity in female mammals. *Hormones and Behavior*, *7*, 105–138.

Bekoff, M. (1972). The development of social interaction, play and metacommunication in mammals. *Quarterly Review of Biology*, *47*, 412–434.

Bekoff, M. (1995). Cognitive ethology and the explanation of nonhuman animal behavior. In H.L. Roitblat, & J.-A. Meyer (Eds.), *Comparative appproaches to cognitive science*, (pp.119–150). Cambridge, MA: MIT Press.

Bérard, J.D., Nurberg, P., Epplen, J.T., & Schmidtke, J. (1993). Male rank, reproductive behavior and reproductive success in free-ranging rhesus macaques. *Primates*, *34*, 481–489.

Bercovitch, F.B. (1988). Coalitions, cooperation and reproductive tactics among adult male baboons. *Animal Behaviour*, *36*, 1198–1209.

Bercovitch, F.B. (1995). Female cooperation, consortship maintenance and mating success in savanna baboons. *Animal Behaviour*, *50*, 137–149.

Bernstein, I.S. (1970). Primate status hierarchies. In L.A. Rosenblum (Ed.), *Primate behavior developments in field and laboratory research* (Vol. 1, pp.71–109). New York: Academic Press.

Bernstein, I.S. (1981). Dominance: The baby and the bathwater, *Behavioural and Brain Sciences*, *4*, 419–457.

Bertrand, M. (1969). *The behavioural repertoire of the stumptail macaque: A descriptive and comparative study*. Basel: S. Karger.

Biben, M., & Symmes, D. (1986). Play vocalizations of squirrel monkeys (*Saimiri sciureus*). *Folia Primatologica*, *46*, 173–182.

Bickerton, D. (1990). *Language and species*. Chicago: University of Chicago Press.

Boehm, C. (1989). Methods for isolating chimpanzee vocal communication. In P.G. Heltne, & L.A. Marquadt (Eds.), *Understanding chimpanzees*, (pp.38–59). Cambridge, MA: Harvard University Press (in cooperation with the Chicago Academy of Sciences).

Boesch, C. (1991). Teaching among wild chimpanzees. *Animal Behaviour, 41*, 530–532.

Boinski, S. (1991). The coordination of spatial position: A field study of adult female squirrel monkeys. *Animal Behaviour, 41*, 89–102.

Boinski, S. (1992). Ecological and social factors affecting the vocal behaviour of adult female squirrel monkeys. *Ethology, 92*, 316–330.

Boinski, S. (1993). Vocal coordination of troop movement among white-faced capuchin monkeys, *Cebus capucinus*. *American Journal of Primatology, 30*, 85–100.

Boinski, S., & Mitchell, C.L. (1995). Wild squirrel monkey (*Saimiri Sciureus*) "caregiver" calls: Contexts and acoustic structure. *American Journal of Primatology, 35*, 129–137.

Boorman, S.A., & Levitt, P.R. (1980). The comparative evolutionary biology of social behaviour. *Annual Review of Sociology, 6*, 213–237.

Boysen, S.T., & Berntson, G.G. (1989). Conspecific recognition in the chimpanzee (*Pan troglydytes*): Cardiac responses to significant others. *Journal of Comparative Psychology, 103*, 215–220.

Bruner, J.S. (1975). From communication to language: A psychological perspective. *Cognition, 3*, 255–287.

Buck, R. (1991). Social factors in facial display and communication: A reply to Chovil and others. *Journal of Nonverbal Behaviour, 15*, 155–161.

Burghardt, G.M. (1985). Animal awareness. Current perceptions and historical perspective, *American Psychologist, 40*, 905–919.

Buss, D.M. (1995). Evolutionary psychology: A new paradigm for psychological science, *Psychological Inquiry, 6*, 1–30.

Busse, C. (1984). Triadic interactions among male and infant chacma baboons. In D.M. Taub (Ed.), *Primate paternalism*, (pp.186–212). New York: Van Norstrand Reinhold.

Busse, C., & Hamilton, W.J. (1981). Infant carrying by male chacma baboons. *Science, 212*, 1281–1283.

Butovskaya, M., Kozintsev, A., & Welker, C. (1996). Conflict and reconciliation in two goups of crab-eating monkeys differing in social status by birth. *Primates, 37*, 261–270.

Butterworth, G. (1995). Factors in visual attention eliciting manual pointing in human infancy. In H.L. Roitblat, & J-A. Meyer (Eds.), *Comparative Appproaches to Cognitive Science*, (pp.329–338). Cambridge, MA: MIT Press.

Byrne, R.W. (1982). Primate vocalizations: Structural and functional approaches to understanding. *Behaviour, 80*, 241–258.

Byrne, R.W. (1995). *The thinking ape. Evolutionary origins of intelligence*. Oxford: Oxford University Press.

Byrne, R.W. (1997). The technical intelligence hypothesis: An additional evolutionary stimulus to intelligence. In A. Whiten, & R.W. Byrne (Eds.), *Machiavellian Intelligence II*, (pp.289–311). Cambridge: Cambridge University Press.

Byrne, R.W., & Whiten, A. (1985). Tactical deception of familiar individuals in baboons (*Papio ursinus*). *Animal Behavior, 33*, 669–673.

Byrne, R.W., & Whiten, A. (1987). The thinking primate's guide to deception. *New Scientist, 116*, 54–57.

Byrne, R.W., & Whiten, A. (1988). *Machiavellian intelligence. Social expertise and the evolution of intellect in monkeys, apes and humans*. Oxford: Clarendon Press.

Call, J., & Rachat P. (1996). Liquid conservation in orang-utans (*Pongo pygmaeus*) and humans (*Homo sapiens*): Individual differences and perceptual strategies. *Journal of Comparative Psychology, 110*, 219–232.

Call, J., & Tomasello, T. (1994). The production and comprehension of referential pointing by orang-utans (*Pongo pygmaeus*). *Journal of Comparative Psychology, 108*, 307–317.

Callan, H. (1984). The imagery of choice in sociobiology. *Man, 19*, 404–420.

Carpenter, C.R. (1934). A field study of the behavior and social relations of the howling monkeys (*Alouatta palliata*). *Comparative Psychology Monographs, 10*, (2), Serial No. 48, 1–168.

Carpenter, C.R. (1940). A field study in Siam of the behaviour and social relations of the gibbon (*Hylobates lar*). *Comparative Psychology Monographs, 16,* (5), Serial No. 84, 1–206.
Carpenter, C.R. (1942). Sexual behaviour of free-ranging rhesus monkeys (*Macaca mulatta*). I. Specimens, procedures and behavioural characteristics of estrus. *Journal of Comparative Psychology, 33,* 113–142.
Carpenter, C.R. (1945). Concepts and problems of primate sociometry. *Sociometry, 8,* 56–61.
Chadwick-Jones, J.K. (1989). Presenting and mounting in nonhuman primates; New theoretical developments. *Journal of Biological and Social Structures, 12,* 319–333.
Chadwick-Jones, J.K. (1991). The social contingency model and olive baboons. *International Journal of Primatology, 12,* 145–161.
Chaffin, C.L., Friedlen, K., & de Waal, F.B.M. (1995). Dominance style of Japanese macaques compared with rhesus and stumptail macaques. *American Journal of Primatology, 35,* 103–116.
Chalmeau, R., & Gallo, A. (1996). What chimpanzees (*Pan troglodytes*) learn in a cooperative task. *Primates, 37,* 39–49.
Chalmeau, R., Lardeux, K., Brandibas, P., & Gallo, A. (1997). Cooperative problem solving by orangutans (*Pongo pygmaeus*). *International Journal of Primatology, 18,* 23–34.
Chance, M.R.A. (1956). Social structure of a colony of *Macaca mulatta. British Journal of Animal Behaviour, 4,* 1–13.
Chapais, B. (1996). Competing through cooperation in nonhuman primates: Developmental aspects of matrilineal dominance. *International Journal of Behavioral Development, 19,* 7–23.
Chapman, C.A., & Lefebre, L. (1990). Manipulating foraging group size: Spider monkeys food calls at fruiting trees. *Animal Behaviour, 39,* 891–896.
Chapman, C.A., & Weary, D.M. (1990). Variability in spider monkeys' vocalizations may provide basis for individual recognition. *American Journal of Primatology, 22,* 279–284.
Chapman, C.A., Wrangham, R.W., & Chapman, L.J. (1995). Ecological constraints on group size: An analysis of spider monkey and chimpanzee subgroups. *Behavioural Ecology and Sociobiology, 36,* 59–70.
Cheney, D.L., & Seyfarth, R.M. (1980). Vocal recognition in free-ranging vervet monkeys. *Animal Behaviour, 28,* 362–367.
Cheney, D.L., & Seyfarth, R.M. (1982). How vervet monkeys perceive their grunts: Field playback experiments. *Animal Behaviour, 30,* 739–751.
Cheney, D.L., & Seyfarth, R.M. (1985). Vervet monkey alarm calls: Manipulation through shared information? *Behaviour, 93,* 150–166.
Cheney, D.L., & Seyfarth, R.M. (1986). The recognition of social alliances among vervet monkeys. *Animal Behaviour, 34,* 1722–1731.
Cheney, D.L., & Seyfarth, R.M. (1990a). *How monkeys see the world: Inside the mind of another species.* Chicago: University of Chicago Press.
Cheney, D.L., & Seyfarth, R.M. (1990b). The representation of social relations by monkeys. *Cognition, 37,* 167–196.
Cheney, D.L., & Seyfarth, R.M. (1991). Reading minds or reading behaviour? Tests for a theory of mind in monkeys. In A. Whiten (Ed.), *Natural theories of mind. Evolution, development and simulation of everyday mindreading,* (pp.175–194). Oxford: Basil Blackwell.
Cheney, D.L., Seyfarth, R.M., & Palombit, R. (1996). The functions and mechanisms underlying baboon "contact" barks. *Animal Behaviour,* 52, 507–518.
Cheney, D.L., Seyfarth, R.M., & Silk, J.B. (1995a). The role of grunts in reconciling opponents and facilitating interactions among adult female baboons. *Animal Behaviour, 50,* 249–257.
Cheney, D.L., Seyfarth, R.M., & Silk, J.B. (1995b). The responses of female baboons (*Papio cynocephalus ursinus*) to anomalous social interactions: Evidence of causal reasoning? *Journal of Comparative Psychology, 109,* 134–141.
Chiarello, A.G. (1995). Grooming in six brown howler monkeys (*Alouatta fusca*). *American Journal of Primatology, 35,* 73–81.

Chovil, N. (1991). Social determinants of facial displays. *Journal of Nonverbal Behaviour, 15*, 141–154.

Chovil, N., & Fridlund, A.J. (1991). Why emotionality cannot equal sociality: Reply to Buck. *Journal of Nonverbal Behaviour, 15*, 163–167.

Clark, A.P. (1993). Rank differences in the vocal production of Kibale Forest chimpanzees as a function of social context. *American Journal of Primatology, 31*, 159–179.

Clark, A.P., & Wrangham, R.W. (1993). Acoustic analysis of wild chimpanzees pant hoots: Do Kibale Forest chimpanzees have an acoustically distinct food arrival pant hoot? *American Journal of Primatology, 31*, 99–109.

Clutton-Brock, T.H., Guinness, F.E., & Albon, S.D. (1982). *Red deer: The behaviour and ecology of two sexes*. Chicago: University of Chicago Press.

Clutton-Brock, T.H., & Harvey, P.H. (1976). Evolutionary rules and primate societies. In P.P.G. Bateson & R.A. Hinde (Eds.), *Growing Points in Ethology*, (pp.195–237). Cambridge: Cambridge University Press.

Colmenares, F. (1990). Greeting behavior in male baboons I: Communication, reciprocity and symmetry. *Behaviour, 113*, 81–116.

Colmenares, F. (1991). Greeting behaviour between male baboons: Oestrus females, rivalry and negotiation. *Animal Behaviour, 41*, 49–60.

Conroy, G.C. (1990). *Primate evolution*. New York: W.H. Norton.

Cooper, J., & Fazio, R.H. (1984). A new look at dissonance theory. In L. Berkowitz (Ed.), *Advances in Experimental Social Psychology, Vol 17* (pp.229–266). New York: Academic Press.

Cords, M. (1992). Post-conflict reunions and reconciliation in pig-tailed macaques. *Animal Behaviour, 44*, 57–61.

Cords, M. (1993a). On operationally defining reconciliation. *American Journal of Primatology, 29*, 255–267.

Cords, M. (1993b). Grooming and language as cohesion mechanisms: Choosing the right data. *Behavioural and Brain Sciences, 16*, 697–698.

Cords, M. (1995). Predator vigilance costs of allogrooming in wild blue monkeys. *Behaviour, 132*, 559–569.

Cords, M. (1997). Friendships, alliances, reciprocity and repair. In A. Whiten, & R.W. Byrne (Eds.), *Machiavellian Intelligence II*, (pp.24–49). Cambridge: Cambridge University Press.

Cords, M., & Aureli, F. (1993). Coping with aggression by juvenile long-tailed macaques (*Macaca fasicularis*). In M.E. Pereira, & L.A. Fairbanks (Eds.), *Juvenile primates: Life history, development and behaviour*, (pp.271–284). New York: Oxford University Press.

Cords, M., & Aureli, F. (1996). Reasons for reconciling. *Evolutionary Anthropology, 5*, 42–45.

Cords, M., & Thurnheer, S. (1993). Reconciling with valuable partners by long-tailed macaques. *Ethology, 93*, 315–325.

Cosmides, L., & Tooby, J. (1989). Evolutionary psychology and the generation of culture. Part II. Case study: A computational theory of social exchange. *Ethology and Sociobiology, 10*, 51–97.

Cosmides, L., & Tooby, J. (1993). The lords of many domains. *Times Higher Educational Supplement, Vol 1077, Perspectives Special: Evolution and Human Sciences*, ii–iii.

Custance, D.M., & Bard, K.A. (1994). The comparative and developmental study of self-recognition and imitation: The importance of social factors. In S.T. Parker, R.W. Mitchell, & M. Boccia (Eds.), *Self-awareness in animals and humans: Developmental perspectives* (pp.207–226). Cambridge, UK: Cambridge University Press.

Darwin, C. (1871). *The descent of man and selection in relation to sex*. London: J. Murray.

Darwin, C. (1876). Sexual selection in relation to monkeys. *Nature, 15*, 18–19.

Dasser, V. (1988a). A social concept in Java monkeys. *Animal Behaviour, 36*, 225–230.

Dasser, V. (1988b). Mapping social concepts in monkeys. In R.W. Byrne, & A. Whiten (Eds.), *Machiavellian Intelligence*, (pp.85–93). Oxford: Oxford University Press.

Dawkins, M.S. (1995). *Unravelling animal behaviour* (2nd Edn). Harlow, UK: Longman.

Deag, J.M., & Crook, J.H. (1971). Social behaviour and "agonistic buffering" in the wild Barbary macaque (*Macaca sylvana L.*). *Folia Primatologica, 15*, 183–200.

Delacour, J. (1995). A model of the brain and memory system. In H.L. Roitblat, & J.A Meyer (Eds.), *Comparative approaches to cognitive science*, (pp.305–327). Cambridge, MA: MIT Press.

de Blois, S.T., & Novak, M.A. (1994). Object permanence in rhesus monkeys (*Macaca mulatta*). *Journal of Comparative Psychology, 108*, 318–327.

de Waal, F.B.M. (1977). The organization of agonistic relations within two captive groups of Java monkeys (*Macaca fascicularis*). *Zeitschrift für Tierpsychologie, 44*, 225–282.

de Waal, F.B.M. (1978). Exploitative and familiarity—dependent support strategies in a colony of semi-free living chimpanzees, *Behaviour, 66*, 268–312.

de Waal, F.B.M. (1982). *Chimpanzee politics: Power and sex among apes*. London: Jonathan Cape.

de Waal, F.B.M. (1986a). Deception in the natural communication of chimpanzees. In R.W. Mitchell, & N.S. Thompson (Eds.), *Deception: Perspectives on human and nonhuman deceit*, (pp.221–244). New York: State University of New York Press.

de Waal, F.B.M. (1986b). The integration of dominance and social bonding in primates. *Quarterly Review of Biology, 61*, 459–479.

de Waal, F.B.M. (1987). Dynamics of social relationships. In B.B. Smuts, D.L. Cheney, R.M. Seyfarth, R.W. Wrangham, & T.T. Struhsaker (Eds.), *Primate societies*, (pp.421–429). Chicago: University of Chicago Press.

de Waal, F.B.M. (1988). The comparative repertoire of captive bonobos (*Pan paniscus*) compared to that of chimpanzees. *Behaviour, 106*, 183–251.

de Waal, F.B.M. (1989a). *Peacemaking among primates*. Cambridge, MA: Harvard University Press.

de Waal, F.B.M. (1989b). Food-sharing and reciprocal obligations among chimpanzees. *Journal of Human Evolution, 18*, 433–459.

de Waal, F.B.M. (1989c). Behavioural contrasts between bonobo and chimpanzee. In P.G. Heltne, & L.A. Marquadt (Eds.), *Understanding chimpanzees*, (pp.154–175). Cambridge, MA: Harvard University Press (in cooperation with the Chicago Academy of Sciences).

de Waal, F.B.M. (1992). Coalitions as part of reciprocal relations in the Arnhem chimpanzee colony. In A.H. Harcourt, & F.B.M. de Waal (Eds.), *Coalitions and alliances in humans and other animals*, (pp.233–257). Oxford: Oxford University Press.

de Waal, F.B.M. (1996a). *Good natured*. Cambridge, MA: Harvard University Press.

de Waal, F.B.M. (1996b). Macaque social culture: Development and perpetuation of affiliative networks. *Journal of Comparative Psychology, 110*, 147–154.

de Waal, F.B.M., & Aureli, F. (1996). Consolation, reconciliation and a possible cognitive difference between macaques and chimpanzees. In A.E. Russon, K.A. Bard, & S.T. Parker (Eds.), *The minds of great apes*. Cambridge: Cambridge University Press.

de Waal, F.B.M., & Luttrell, L.M. (1985). The formal hierarchy of rhesus macaques: An investigation of the bared-teeth display. *American Journal of Primatology, 9*, 73–85.

de Waal, F.B.M., & Luttrell, L.M. (1988). Mechanisms of social reciprocity in three primate species: Symmetrical relationship characteristics or cognition? *Ethology and Sociobiology, 9*, 101–118.

de Waal, F.B.M., & van Hooff, J.A.R.A.M (1981). Side-directed communication and agonistic interactions in chimpanzees. *Behaviour, 77*, 164–198.

de Waal, F.B.M., & van Roosmalen, A. (1979). Reconciliation and consolation among chimpanzees. *Behavioural Ecology and Sociobiology, 5*, 55–66.

Digby, L.J. (1995). Social organization in a wild population of *Callithrix jacchus*: II. Intragroup Social Behaviour. *Primates, 35*, 361–375.

Domjan, M., & Purdy, J.E. (1995). Animal research in psychology: More than meets the eye of the general psychology student. *American Psychologist, 50*, 496–503.

Donaldson, M. (1978). *Children's minds*. London: Fontana.

Dunbar, R.I.M. (1980). Determinants and evolutionary consequences of dominance among female gelada baboons. *Behavioural Ecology and Sociobiology, 7*, 253–265.

Dunbar, R.I.M. (1984). *Reproductive decisions. An economic analysis of gelada baboon social strategies*. Princeton, NJ: Princeton University Press.

Dunbar, R.I.M. (1988). *Primate social systems*. London: Croom Helm.

Dunbar, R.I.M. (1993). Coevolution of neocortical size, group size and language in humans. *Behavioral and Brain Sciences, 16*, 681–735.

Dunbar, R.I.M., & Sharman, M. (1984). Is social grooming altruistic? *Zeitschrift für Tierpsychologie, 64*, 163–173.

Eddy, T.J., Gallup, G.G., & Povinelli, D.J. (1996). Age differences in the ability of chimpanzees to distinguish mirror-images of self from video images of others. *Journal of Comparative Psychology, 110*, 38–44.

Eibl-Eibesfeldt, I. (1972). Similarities and differences between culture in expressive movements. In R.A. Hinde (Ed.), *Non-verbal communication*, (pp.297–314). Cambridge: Cambridge University Press.

Ekman, P. (Ed.) (1982). *Emotions in the human face* (2nd Edn). Cambridge: Cambridge University Press.

Ekman, P. (1992). Facial expressions of emotion: New findings, new questions. *Psychological Science, 3*, 34–38.

Ekman, P. (1994). Strong evidence for universals in facial expressions: A reply to Russell's mistaken critique. *Psychological Bulletin, 115*, 268–287.

Ekman, P., Davidson, R.J., & Friesen, W.V. (1990). The Duchenne smile: Emotional expression and brain physiology II. *Journal of Personality and Social Psychology, 58*, 342–353.

Ekman, P., & Friesen, W.V. (1969). The repertoire of nonverbal behavior: Categories, origins, usage, and coding, *Semiotica, 1, 49–98.*

Ekman, P., & Friesen, W.V. (1975). *Unmasking the face: A guide to recognizing emotions from facial cues*. Englewood Cliffs, NY: Prentice-Hall.

Ekman, P., & Friesen, W.V. (1978). *Facial action coding system*. Palo Alto, CA: Consulting Psychologists Press.

Ekman, P., & Friesen, W.V. (1982a). Measuring facial movement with the Facial Action Coding System. In P. Ekman (Ed.), *Emotion in the human face* (2nd Edn). Cambridge: Cambridge University Press.

Ekman, P., & Friesen, W.V. (1982b). Felt, false and miserable smiles. *Journal of Nonverbal Behaviour, 6*, 238–252.

Etienne, A.S. (1989). The application of the Piagetian stage concept to comparative research. *Behavioral and Brain Sciences, 12*, 595.

Estrada, A., Estrada, R., & Ervin, F. (1977). Establishment of a free-ranging colony of stumptail macaques (*Macaca arctoides*): Social relations 1. *Primates, 18*, 647–676.

Fady, J.C. (1990). Les différents domaines de recherche en primatologie: Leur importance relative et leur evolution depuis 1980. *Sciences et Techniques de l'Animal Laboratorie, 15*, 165–177.

Fairbanks, L.A. (1980). Relationships among adult females in captive vervet monkeys: Testing a model of rank-related attractiveness. *Animal Behaviour, 28*(3), 853–859.

Fausto-Sterling, A. (1985). *Myths of gender: Biological theories about women and men*. New York: Basic Books.

Fedigan, L.M. (1982). *Primate paradigms: Sex roles and social bonds*. Montréal: Eden Press.

Fedigan, L.M. (1997). Changing views of female life histories. In M.E. Morbeck, A. Galloway, & A.L. Zihlman (Eds.), *The evolving female*, (pp.15–26). Princeton NJ: Princeton University Press.

Fernandez Dols, J.M., & Ruiz Belda, M-A. (1997). Spontaneous facial behaviour during intense emotional episodes: Artistic truth and optical truth. In J.A. Russell, & J.M. Fernandez Dols (Eds.), *The psychology of facial expression*, (pp.255–274). Cambridge: Cambridge University Press.

Feyereisen, P., & de Lannoy, J-D. (1991). *Gestures and speech. Psychological investigation*. Cambridge: Cambridge University Press.

Filion, C.M., Washburn, D.A., & Gulledge, J.P. (1996). Can monkeys (*Macaca mulatta*) represent invisible displacement? *Journal of Comparative Psychology*, *110*, 386–395.

Fillion, K. (1997). *Lip service. The myth of female virtue in love, sex and friendship*. London: Pandora (HarperCollins).

Fisher, J.A. (1996). The myth of anthropomorphism. In M. Bekoff & D. Jamieson (Eds.), *Readings in animal cognition*. Cambridge, MA: MIT Press.

Flavell, J.H. (1993). *Cognitive development* (3rd Edn). Englewood Cliffs, NJ: Prentice Hall.

Foa, U.G., Converse J., Tornblom, K.Y., & Foa E.B. (Eds.) (1993). *Resource theory. Explorations and applications*. San Diego, CA: Academic Press.

Fragaszy, D.M. (1981). Comparative performance in discrimination learning tasks in two New World primates. (*Salmiri sciureus* and *Callicebus moloch*). *American Journal of Primatology*, *2*, 191–200.

Frank, M.G., Ekman, P., & Friesen, W.V. (1993). Behavioural markers and recognizability of the smile of enjoyment. *Journal of Personality and Social Psychology*, *64*, 83–93.

Fridlund, A.J. (1994). *Human facial expression: An evolutionary view*. San Diego, CA: Academic Press.

Frye, D. (1991). The origins of intention in infancy. In D. Frye, & C. Moore (Eds.), *Children's theories of mind: Mental states and social understanding*, (pp.15–38). Hillsdale, NJ: Lawrence Erlbaum.

Furth, H. (1992). Theory of society, yes, theory of mind, no. *Behavioral and Brain Sciences*, *15*, 155–156.

Galdikas, B.M.F. (1978). *Orangutan adaptation at Tanjung Puting Reserve, Central Borneo*. Ph.D. Dissertation, Los Angeles: University of California.

Galdikas, B.M.F. (1984). Adult female sociality among wild orangutans at Tanjing Puting Reserve. In M.F. Small (Ed.) *Female primates: Studies by women primatologists*, (pp.217–235). New York: Alan R. Liss.

Galdikas, B.M.F. (1988). Orangutan diet, range and activity at Tanjun Puting, Central Borneo. *International Journal of Primatology*, *9*, 1–31.

Galdikas, B.M.F. (1992). Social and nonsocial intelligence in orangutans. *Behavioral and Brain Sciences*, *15*, 156–157.

Gale, A., Kingsley, E., Brookes, S., & Smith, D. (1978). Cortical arousal and social intimacy in the human female under different conditions of eye contact. *Behavioural Processes*, *3*, 271–275.

Gallup, G.G. (1970). Chimpanzees: Self-recognition. *Science*, *167*, 86–87.

Gallup, G.G. (1977). Self-recognition in primates: A comparative approach to the bidirectional properties of consciousness. *American Psychologist*, *32*, 329–338.

Gallup, G.G. (1982). Self-awareness and the emergence of mind in primates. *American Journal of Primatology*, *2*, 237–248.

Gallup, G.G. (1983). Toward a comparative psychology of mind. In R.L. Mellgren (Ed.), *Animal cognition and behavior*, (pp.473–510). Amsterdam: North Holland.

Gallup, G.G., & Povinelli, D.J. (1993). Mirror, mirror on the wall which is the most heuristic theory of them all? A response to Mitchell. *New Ideas in Psychology*, *11*, 327–335.

Gallup, G.G., & Suarez, S.D. (1991). Social responding to mirrors in rhesus monkeys (*Macaca mulatta*): Effects of temporary mirror removal. *Journal of Comparative Psychology*, *105*, 376–379.

Gartlan, J.S. (1968). Structure and function in primate society. *Folia Primatologica*, *8*, 89–120.

Gautier, J.P., & Gautier, A. (1977). Communication in Old World monkeys. In T.A. Seboek (Ed.), *How animals communicate*, (pp.890–964). Bloomington: Indiana University Press.

Ghiglieri, M.P. (1987). Sociobiology of the great apes and the hominid ancestor. *Journal of Human Evolution*, *16*, 319–357.

Ghiglieri, M.P. (1989). Hominoid sociobiology and hominoid social evolution. In P.G. Heltne, & L.A. Marquadt (Eds.), *Understanding chimpanzees*, (pp.370–379). Cambridge, MA: Harvard University Press (in cooperation with the Chicago Academy of Sciences).

Ginsburg, G.P. (1997). Faces: An epilogue and reconceptualization. In J.A. Russell, & J.M. Fernandez Dols, (Eds.), *The psychology of facial expression*, (pp.349–380). Cambridge: Cambridge University Press.

Goldizen, A.W. (1987). Tamarins and marmosets: Communal care of offspring. In B.B. Smuts, D.L. Cheney, R.M. Seyfarth, R.W. Wrangham, & T.T. Struhsaker (Eds.), *Primate societies* (pp.34–43). Chicago: University of Chicago Press.

Gomez, J.C. (1991). Visual behaviour as a window for reading the mind of others in primates. In A. Whiten (Ed.), *Natural theories of mind. Evolution, development and simulation of everyday mindreading*, (pp.195–207). Oxford: Basil Blackwell.

Gomez, J.C. (1996). Ostensive behavior in great apes; the role of eye-contact. In A.E. Russon, K.A. Bard, & S.T. Parker (Eds.), *Reaching into thought. The minds of the great apes*, (pp.131–151). Cambridge: Cambridge University Press.

Goodall, J. (1968). The behaviour of free-living chimpanzees in the Gombe Stream Reserve. *Animal Behaviour Monographs*, *1*, 161–311.

Goodall, J. (1972). A preliminary report on expressive movements and communication in the Gombe Stream chimpanzees. In P. Dolhinow (Ed.), *Primate patterns*, (pp.25–84). New York: Holt, Rhinehart & Winston.

Goodall, J. (1986). *The chimpanzees of Gombe: Patterns of behavior*. Cambridge, MA: Belknap Press of Harvard University Press.

Goodenough, J., McGuire, B., & Wallace, R.A. (1993). *Perspectives on animal behaviour*. New York: John Wiley & Sons.

Goosen, C. (1980). *On grooming in Old World monkeys*. Delft: Meinema.

Gould, S.J. (1977). *Ontogeny and phylogeny*. Cambridge, MA: Harvard University Press.

Gouzoules, H. (1992). Is this best of all possible worlds? *Behavioral and Brain Sciences*, *15*, 135.

Gouzoules, H., Gouzoules, S., & Ashley, J. (1995). Representational signalling in non-human primate communication. In E. Zimmermann, J.D. Newman, & Jurgens, U. (Eds.), *Current topics in primate vocal communication*, (pp.235–252). New York: Plenum.

Gouzoules, S., Gouzoules, H., & Marler, P. (1984). Rhesus monkeys (*Macaca mulatta*) screams: Representational signalling in the recruitment of agonistic aid. *Animal Behaviour*, *32*, 182–193.

Green, S. (1975). Variation of vocal pattern with social situation in the Japanese monkey (*Macaca fuscata*): A field study. In L.A. Rosenblum (Ed.), *Primate behaviour*, Vol 4, (pp.2–102). New York: Academic Press.

Griffin, D.R. (1981). *The question of animal awareness. Evolutionary continuity of mental experience.* (Rev Edn). Los Altos, CA: William Kaufmann.

Guilford, T., & Dawkins, M.S. (1991). Receiver psychology and the evolution of animal signals. *Animal Behaviour*, *42*, 1–14.

Gruen, N.L. (1996). Gendered knowledge. Examining influences on scientific and ethological knowledge. In M. Bekoff, & D. Jamieson (Eds.), *Readings in animal cognition*, (pp.17–27). Cambridge, MA: MIT Press.

Hall, K.R.L., & DeVore, I. (1965). Baboon social behavior. In I. DeVore (Ed.), *Primate behavior. Field studies of monkeys and apes*, (pp.53–110). New York: Holt, Rinehart & Winston.

Hamilton, W.D. (1984). Significance of paternal investment by primates to the evolution of male–female associations. In D.M. Taub (Ed.), *Primate paternalism*, (pp.309–335). New York: Van Nostrand Reinhold.

Hanby, J.P. (1972). The sociosexual nature of mounting and related behaviours in a confined troop of Japanese macaques (*Macaca fuscata*). University of Oregon, Ph.D. Thesis.

Haraway, D. (1989). *Primate visions*. London: Routledge.

Harcourt, A.H. (1979). Contrasts between male relationships in wild gorilla groups. *Behavioural Ecology and Sociobiology*, *5*, 39–49.

Harcourt, A.H. (1988). Alliances in contests and social intelligence. In R.W. Byrne & A. Whiten (Eds.), *Machiavellian intelligence. Social expertise and the evolution of intellect in monkeys, apes and humans*, (pp.132–152). Oxford: Clarendon Press.

Harcourt, A.H. (1992). Coalitions and alliances: Are primates more complex than non-primates? In A.H. Harcourt & F.B.M. de Waal (Eds.), *Coalitions and alliances in humans and other animals*, (pp.445–471). Oxford: Oxford University Press.

Harcourt, A.H., & Stewart, K.J. (1987). The influence of help in contests or dominance rank in primates: Hints from gorillas. *Animal Behaviour*, *35*, 182–190.

Harcourt, A.H., & Stewart, K.J. (1989). Functions of alliances in contests within wild gorilla groups. *Behaviour*, *109* (3/4), 176–190.

Harcourt, A.H., Stewart, K.J., & Hauser, M. (1993). Functions of wild gorillas "close" calls. I. Repertoire, context and interspecific comparison. *Behaviour*, *124*, 89–121.

Harcourt, A.H., & de Waal F.B.M. (Eds.) (1992). *Coalitions and alliances in humans and other animals*. Oxford: Oxford University Press.

Hargie, O., Saunders, C., & Dickson, D. (1994). *Social skills in interpersonal communication* (3rd Edn). London: Routledge.

Harré, R. (1984). Vocabularies and theories. In R. Harré & V. Reynolds (Eds.), *The meaning of primate signals*, (pp.90–110). Cambridge: Cambridge University Press.

Harris, R. (1984). Must monkeys mean? In R. Harré & V. Reynolds (Eds.), *The meaning of primate signals*, (pp.116–137). Cambridge: Cambridge University Press.

Hart, D., & Fegley, S. (1994). Social imitation and the emergence of a mental model of self. In S.T. Parker, R.W. Mitchell, & M. Boccia (Eds.), *Self-awareness in animals and humans. Developmental perspectives*, (pp.149–165). Cambridge: University of Cambridge.

Hasegawa, T. (1989). Sexual behaviour of immigrant and resident female chimpanzees at Mahale. In P.G. Heltne & L.A. Marquadt (Eds.), *Understanding chimpanzees*, (pp.90–103). Cambridge, MA: Harvard University Press (in cooperation with the Chicago Academy of Sciences).

Hassett, J. (1984). *Psychology in perspective*. New York: Harper & Row.

Hauser, M., Teixidor, P., Field, L., & Flaherty, R. (1993). Food-elicited calls in chimpanzees: Effects of food quantity and divisibility. *Animal Behaviour*, *45*, 817–19.

Hausfater, G., & Takacs, D. (1987). Structure and function of hindquarter presentations in yellow baboons (*Papio cynocephalus*). *Ethology*, 74, 297–319.

Hebb, D.O. (1946). Emotion in man and animal: An analysis of the intuitive processes of recognition. *Psychological Review*, *53*, 88–106.

Hemelrijk, C.K. (1990a). A matrix partial correlation test used in investigations of reciprocity and other social interaction patterns at group level. *Journal of Theoretical Biology*, *143*, 405–420.

Hemelrijk, C.K. (1990b). Models of and tests for, reciprocity, unidirectionality and other social interaction patterns at a group level. *Animal Behaviour*, *39*, 1013–1029.

Hemelrijk, C.K., & Ek, A. (1991). Reciprocity and interchange of grooming and "support" in captive chimpanzees. *Animal Behaviour*, *41*, 923–935.

Herschberger, R. (1970). *Adam's rib*. New York: Harper & Row.

Heyes, C.M. (1994). Reflections on self-recognition in primates. *Animal Behaviour*, *47*, 909–919.

Hill, D.A., & Okayasu, N. (1995). Absence of "youngest ascendency" in the dominance relations of sisters in wild Japanese macaques (*Macaca fuscata yakin*). *Behaviour*, *32*, 367–379.

Hinde, R.A. (1982). *Ethology. Its nature and relations with other sciences*. Oxford: Oxford University Press.

Hinde, R.A. (1983). *Primate social relationships: An integrated approach.* Oxford: Blackwell Scientific.

Hinde, R.A. (1985). Expression and negotiation. In G. Zivin (Ed.), *The development of expressive behaviour: Biology–environment interactions*, (pp.103–116). New York: Academic Press.

Hinde, R.A. (1997). *Relationships*. Hove, UK: Psychology Press.

Hinde, R.A., & Datta, S. (1981). Dominance: An intervening variable. *Behavioral and Brain Sciences, 4*, 442.

Hinde, R.A., & Rowell, T.E. (1962). Communication by postures and facial expressions in the rhesus monkey (*Macaca mulatta*). *Proceedings of Zoological Society of London, 138*, 1–21.

Hinde, R.A., & Stevenson-Hinde, J. (1976). Towards understanding relationships: Dynamic stability. In P.P.G. Bateson & R.A. Hinde (Eds.), *Growing points in ethology*, (pp.451–479). Cambridge: Cambridge University Press.

Hohmann, G., & Fruth, B. (1995). Loud calls in great apes: Sex differences and social correlates. In E. Zimmermann, J.D. Newman, & U. Jurgens (Eds.), *Current topics in primate vocal communication*, (pp.161–184). New York: Plenum.

Holmberg, D., & Veroff, J. (1996). Rewriting relationship memories: The effects of courtship and wedding scripts. In G.J.O. Fletcher & J. Fitness (Eds.), *Knowledge structures in close relationships: A social psychological approach*, (pp.345–368) Mahwah, NJ: Lawrence Erlbaum Associates.

Hrdy, S.B. (1974). Male–male competition and infanticide among the langurs (*Presbytis entellus*) of Abu, Rajasthan. *Folia Primatologica, 22*, 19–58.

Hrdy, S.B. (1981). *The woman that never evolved.* Cambridge, MA: Harvard University Press.

Hrdy, S.B., & Whitten, PL. (1987). Patterning of sexual activity. In B.B. Smuts, D.L. Cheney, R.M. Seyfarth, R.W. Wrangham, & T.T. Struhsaker (Eds.), *Primate societies*, (pp.370–384). Chicago: University of Chicago Press.

Huffman, M.A. (1992). Influences of female partner preference on potential reproductive outcome in Japanese macaques. *Folia Primatologica, 59*, 77–88.

Humphrey, N.K. (1976). The social function of intellect. In P.P.G. Bateson & R.A. Hinde (Eds.), *Growing points in ethology*, (pp.303–317). Cambridge: Cambridge University Press.

Humphrey, N.K. (1988). The social function of intellect. In R.W. Byrne & A. Whiten (Eds.), *Machiavellian intelligence. Social expertise and the evolution of intellect in monkeys, apes and humans*, (pp.13–33). Oxford: Clarendon Press.

Idani, G.I. (1995). Function of peering behaviour among bonobos (*Pan paniscus*) at Wamba, Zaire. *Primates, 36*, 377–383.

Inoue-Nakamura, N., & Matsuzawa, T. (1997). Development of stone tool use by wild chimpanzees (*Pan troglodytes*). *Journal of Comparative Psychology, 111*, 159–173.

Itani, J. (1959). Paternal care in the wild Japanese monkey (*Macaca fuscata fuscata*). *Primates, 2*, 61–93.

Izard, C.E. (1977). *Human emotions*. New York: Plenum.

Izard, C.E. (1997). Emotions and facial expressions: A perspective from Differential Emotions Theory. In J.A. Russell & J.M. Fernandez Dols (Eds.), *The psychology of facial expression*, (pp.57–77). Cambridge: Cambridge University Press.

James, W. (1892/1961). *Psychology: The briefer course*. New York: Harper & Row.

Jamieson, D., & Bekoff, M. (1996). On aims and methods of cognitive ethology. In M. Bekoff, & D. Jamieson (Eds.), *Readings in animal cognition*, (pp.65–78). Cambridge, MA: MIT Press.

Jolly, A. (1966). Lemur social behavior and primate intelligence. *Science, 153*, 501–506.

Jolly, A. (1985). *The evolution of primate behavior*. (2nd Edn). New York: MacMillan Press.

Jolly, A. (1988). The evolution of purpose. In R.W Byrne, & A. Whiten (Eds.), *Machiavellian intelligence*, (pp.363–378). Oxford: Oxford University Press.

Judge, P.G. (1991). Dyadic and triadic reconciliation in pigtail macaques (*Macaca nemestrina*). *American Journal of Primatology, 23*, 225–227.

Judge, P.G., & de Waal, F.B.M. (1994). Intergroup grooming relations between Alpha females in a population of free-ranging rhesus monkeys. *Folia Primatologica*, *63*, 63–70.

Kagan, J. (1981). *The second year. The emergence of self-awareness*. Cambridge, MA: Harvard University Press.

Kanazawa, S. (1996). Recognition of facial expressions in a Japanese monkey (*Macaca fuscata*) and humans (*Homo sapiens*). *Primates*, *37*, 25–38.

Kano, T. (1980). Social behaviour of wild pygmy chimpanzees (*Pan paniscus*) of Wamba: A preliminary report. *Journal of Human Evolution*, *9*, 243–260.

Kano, T. (1989). The sexual behaviour of pygmy chimpanzees. In P.G. Heltne, & L.A. Marquadt (Eds.), *Understanding chimpanzees*, (pp.176–183). Cambridge, MA: Harvard University Press (in cooperation with the Chicago Academy of Sciences).

Kano, T. (1992). *The last ape. Chimpanzee behaviour and ecology*. Stanford, CA: Stanford University Press.

Kaplan, J.R. (1978). Fight interference and altruism in rhesus monkeys. *American Journal of Physical Anthropology*, *49*, 241–250.

Kappeler, P., & van Schaik, C.P. (1992). Methodological and evolutionary aspects of reconciliation among primates. *Ethology*, *92*, 51–69.

Kawamura, S. (1958). The matriarchal order in the Minoo-B group—a study on the rank systems of Japanese macaques. *Primates*, *1*, 149–156.

Keating, C.F., & Keating, E.G. (1982). Visual scan patterns of rhesus monkeys viewing faces. *Perception*, *11*, 211–219.

Keating, C.F., & Keating, E.G. (1993). Monkeys and mug shots: cues used by rhesus monkeys (*Macaca mulatta*) to recognize a human face. *Journal of Comparative Psychology*, *107*, 131–139.

Keltner, D., & Buswell, B.N. (1997). Embarrassment: Its distinct form and appeasement functions. *Psychological Bulletin*, *122*, 250–270.

Kennedy, J. (1992). *The new anthropomorphism*. Cambridge: Cambridge University Press.

Kenrick, D.T., & Simpson, J.A. (1997). Why social psychology and evolutionary psychology need one another. In J.A. Simpson, & D.T. Kenrick (Eds.), *Evolutionary Social Psychology*, (pp.1–20). Mahwah, NJ: Lawrence Erlbaum Associates.

Keuster, J., & Paul, A. (1989). Reproductive strategies of subadult Barbary macaque males at Affenburg Salem. In A.E. Rasa, C. Vogel, & E. Voland (Eds.), *The sociobiology of sexual and reproductive strategies*, (pp.93–109). London: Chapman & Hall.

Keverne, E.B. (1976). Sexual receptivity and attractiveness in the female rhesus monkey. In J.S. Rosenblatt, R.A. Hinde, E. Shaw, & C. Beer. (Eds.), *Advances in the Study of Behaviour*, *7*, (pp.155–200). New York: Academic Press.

King, M.-C., & Wilson, A.C. (1975). Evolution at two levels in humans and chimpanzees. *Science*, *188*, 107–116.

Koyama, N. (1985). Playmate relationships among individuals of the Japanese monkey troop in Arashiyama. *Primates*, *26*, 390–406.

Krebs, J.R. (1991). Animal communication: Ideas derived from Tinbergen's activities. In M.S. Dawkins, R.T. Halliday, & R. Dawkins (Eds.), *The Tinbergen legacy*, (pp.60–72). London: Chapman & Hall.

Krebs J.R., & Davies N.B. (Eds.) (1993). *Behavioural ecology. An evolutionary approach*. (2nd Edn). Oxford: Blackwell Scientific.

Krebs, J.R., & Dawkins, R. (1984). Animal signals: Mind reading and manipulation. In J.R. Krebs & N.B. Davies (Eds.), *Behavioural ecology. An evolutionary approach* (2nd Edn), (pp.380–402). Oxford: Blackwell Scientific.

Kummer, H. (1967). Tripartite relations in hamadryas baboons. In S.A. Altmann (Ed.), *Social communication among primates*, (pp.63–71). Chicago: University of Chicago Press.

Kummer, H. (1968). *Social organization of hamadryas baboons. A field study*. Chicago: University of Chicago Press.

Kummer, H. (1975). Rules of dyad and group formation among captive gelada baboons. (*Theropithecus gelada*). In S. Kondo, M. Kawai, A. Ehara, & S. Kawamura (Eds.), *Proceedings from the Symposia of the Fifth Congress of the International Primatological Society*, (pp.129–159). Tokyo: Japan Science Press.

Kummer, H. (1978). On the value of social relationships to nonhuman primates: A heuristic scheme. *Biology and Life. Social Science Information* (London & Beverly Hills), *17*, 687–705.

Kummer, H. (1982). Social knowledge in free-ranging primates. In D.R. Griffin (Ed.), *Animal mind–human mind*, (pp.113–130). Dahlem Konferenzen Berlin: Springer-Verlag.

Kummer, H. (1984). Comment on R. Harré: Vocabularies and theories. In R. Harré & V. Reynolds (Eds.), *The meaning of primate signals*, (pp.106–107). Cambridge: Cambridge University Press.

Kummer, H., Goetz, W., & Angst, W. (1974). Cross-species modifications of social behaviour in baboons. In J.R. Napier & P.H. Napier (Eds.), *Old World monkeys*, (pp.351–363). London: Academic Press.

Kuroda, S. (1980). Social behaviour of the pygmy chimpanzees. *Primates*, *21*, 181–197.

Kuroda, S. (1984). Interaction over food among pygmy chimpanzees. In R.L. Susman (Ed.), *The pygmy chimpanzee. Evolutionary biology and behavior*, (pp.301–324). New York: Plenum Press.

Kuroda, S. (1989). Developmental retardation and behavioral characteristics of pygmy chimpanzees. In P. Heltne & L.A. Marquadt (Eds.), *Understanding chimpanzees*, (pp.184–193). Cambridge, MA: Harvard University Press (in cooperation with the Chicago Academy of Sciences).

Kyes, R.C., & Candland, D.K. (1987). Baboon (*Papio hamadryas*) visual preferences for regions of the face. *Journal of Comparative Psychology*, *101*, 345–348.

Lancaster, J.B. (1975). *Primate behaviour and the emergence of human culture*. New York: Holt, Rhinehart & Winston.

Leavens, D.A., Hopkins, W.D., & Bard, K.A. (1996). Indexical and referential pointing in chimpanzees (*Pan troglodytes*). *Journal of Comparative Psychology*, *110*, 346–353.

Lester, B.M. & Boukydis, C.F.Z. (1992). No language but a cry. In H. Papousek, U. Jurgens, & M. Papousek (Eds.), *Nonverbal vocal communication. Comparative and developmental approaches*, (pp.145–173). Cambridge: Cambridge University Press.

Lieberman, P. (1995). What primate calling tells us about human evolution. In E. Zimmermann, J.D. Newman, & U. Jurgens (Eds.), *Current topics in primate vocal communication*, (pp.273–282). New York: Plenum.

Linnaeus, C. (1758). *Systema Naturae per Regna Tria Naturae, Secundum Classes, Ordines, Genera, Species cum Characteribus, Differentis, Synonymis, Locis.* (10th Edn). Stockholm: Laurentii Salvii.

Loy, J., Argo, B., Nestell, G.-L., Vallett, S., & Wanamaker, G. (1993). A reanalysis of patas monkeys' "grimace and gecker" display and a discussion of their lack of formal dominance. *International Journal of Primatology*, *14*, 879–893.

Macedonia, J.M., & Evans, C.S. (1993). Variations among mammalian alarm call systems and the problem of meaning in animal signals. *Ethology*, *93*, 177–197.

Maestripieri, D. (1993). Vigilance costs of allogrooming in macaque mothers. *American Naturalist*, *141*, 744–753.

Maestripieri, D. (1996a). Gestural communication and its cognitive implications in pigtail macaques (*Macaca Nemestrina*). *Behaviour*, *133*, 997–1022.

Maestripieri, D. (1996b). Primate cognition and the bared-teeth display: A reevaluation of the concept of formal dominance. *Journal of Comparative Psychology*, *110*, 402–405.

Manson, J.H. (1995). Do female rhesus monkeys choose novel mates? *American Journal of Primatology*, *37*, 285–296.

Maple, T.L., & Hoff, M.P. (1982). *Gorilla behavior*. New York: Van Nostrand, Reinhold.

Marler, P. (1976). Social organization, communication and graded signals: The chimpanzee and the gorilla. In P.P.G. Bateson, & R.A. Hinde (Eds.), *Growing points in ethology*, (pp.239–280). Cambridge: Cambridge University Press.

Marler, P. (1977a). Primate vocalization: Affective or symbolic? In G.H. Bourne (Ed.), *Progress in ape research*, (pp.85–96). New York: Academic Press.
Marler, P. (1977b). The evolution of communication. In T.A. Sebeok (Ed.), *How animals communicate*, (pp.45–70). Bloomington: Indiana University Press.
Marler, P. (1992). Functions of arousal and emotion in primate communication: A semiotic approach. In T. Nishida, W.C. McGrew, P. Marler, M. Pickford, & F.B. de Waal (Eds.), *Topics in primatology, Vol.1: Human origins*, (pp.225–233). Tokyo: University of Tokyo Press.
Marler, P., & Tenaza, R. (1977). Signalling behaviour of wild apes with special reference to vocalization. In T.A. Sebeok (Ed.), *How animals communicate*, (pp.965–1033). Bloomington: Indiana University Press.
Martin, R.D. (1974). The biological basis of human behavior. In W.B. Broughton (Ed.), *The biology of brains*. Symposium of the Institute of Biology, No. 21, (pp.215–250). London, Institute of Biology: Distributed by Blackwell, Oxford.
Martin, R.D. (1986). Primates: A definition. In B.A. Wood, L.B. Martin, & P.J. Andrews (Eds.), *Major topics in primate and human evolution*, (pp.1–31). Cambridge: Cambridge University Press.
Martin, R.D. (1990). *Primate origins and evolution. A phylogenetic reconstruction*. Princeton, NJ: Princeton University Press.
Masataka, N. (1989). Motivational referents of contact calls in Japanese macaques. *Ethology*, *80*, 265–273.
Matsumura, S. (1996). Postconflict affiliative contacts among field moor macaques (*Macaca naurus*). *American Journal of Primatology*, *38*, 211–219.
Maxim, P.E. (1976). An internal scale for studying and quantifying social relations in pairs of rhesus monkeys. *Journal of Experimental Psychology (Gen)*, *105*, 123–147.
McFarland, D. (1987). *The Oxford companion to animal behaviour*. Oxford: Oxford University Press.
McGinnis, P.R. (1979). Sexual behaviour in free-living chimpanzees: Consort relationship. In D.A. Hamburg & E.R. McCown (Eds.), *The great apes*, (pp.429–439). Menlo Park, CA: Benjamin/Cummings.
McGrew, W.C. (1992). *Chimpanzee material culture. Implications for human evolution*. Cambridge: Cambridge University Press.
McKenna, J.J. (1982). Biosocial functions of grooming behavior among the common Indian langur monkey (*Presbytis entellus*). *American Journal of Physical Anthropology*, *48*, 503–510.
McNeill, D. (1985). So you think gestures are nonverbal? *Psychological Review*, *92*, 350–371.
Mead, G.H. (1934/1962). *Mind, self and society. From the standpoint of a social behaviorist*. Chicago: University of Chicago Press.
Meltzoff, A.N. (1990). Foundations for developing a concept of self. The role of imitation in relating self to other and the value of social mirroring, social modelling and self practice in infancy. In D. Chichetti, & M. Beeghly (Eds.), *The self in transition. Infancy to childhood*, (pp.139–146). Chicago: University of Chicago Press.
Menzel, C.R. (1997). Primates' knowledge of their natural habitat: As indicated in foraging. In A. Whiten, & R.W. Byrne (Eds.), *Machiavellian Intelligence II*, (pp.207–239). Cambridge: Cambridge University Press.
Miles, H.L.W. (1994). ME CHANTEK: The development of self-awareness in a signing gorila. In S.T. Parker, R.W. Mitchell, & M. Boccia (Eds.), *Self-awareness in animals and humans. Developmental pespectives*, (pp.254–272). Cambridge: University of Cambridge.
Miller, L.C., & Fishkin, S.A. (1997). On the dynamics of human bonding and reproductive success: Seeking windows on the adapted-for human-environment interface. In J.A. Simpson, & D.T. Kenrick (Eds.), *Evolutionary social psychology*, (pp.197–235). Mahwah, NY: Lawrence Erlbaum Associates.
Mitani, J.C. (1985). Mating behavior of male orangutans in the Kutai Reserve, East Kalimantan, Indonesia. *Animal Behaviour*, *33*, 392–402.

Mitani, J.C. (1992). Singing behaviour of male gibbons: Field observations and experiments. In T. Nishida, W. McGrew, P. Marler, M. Pickford, & F. de Waal (Eds.), *Topics in primatology, Volume 1, Human origins*. Tokyo: University of Tokyo Press.

Mitani, J.C. (1996). Comparative studies of African ape vocal behaviour. In W.C. McGrew, L.F. Marchant, & T. Nishida (Eds.), *Great ape societies*, (pp.241–254). Cambridge: Cambridge University Press.

Mitani, J.C., & Brandt, K.L. (1994). Social factors Influence the acoustic variability in the long-distance calls of male chimpanzees. *Ethology*, *96*, 233–252.

Mitani, J.C., Grether, G.F., Rodman, P.S., & Priatna, D. (1996). Association among wild orang-utans: Sociality, passive aggregations or chance? *Animal Behaviour*, *42*, 33–46.

Mitani, J.C., Hasegawa, T., Gros-Luis, J., Marler, P., & Byrne, R. (1992). Dialects in wild chimpanzees? *American Journal of Primatology*, *27*, 233–243.

Mitani, J.C., Gros-Louis, J., & Richards, A.F. (1996). Sexual diomorphism, the operational sex ratio and the intensity of male competition in polygynous primates. *American Naturalist*, *147*, 966–980.

Mitani, J.C., Gros-Louis, J., & Macedonia, J.M. (1996). Selection of acoustic individuality within the vocal repertoire of wild chimpanzees. *International Journal of Primatology*, *17*, 569–583.

Mitani, J.C., & Marler, P. (1989). A phonological analysis of male gibbon singing behavior. *Behaviour*, *109*, 20–45.

Mitchell, R.W. (1993). Mental models of mirror self-recognition: Two theories. *New Ideas in Psychology*, *11*, 295–325.

Mitchell, R.W., & Anderson, J.R. (1993). Discriminative learning of scratching but failure to obtain imitation and self-recognition in a long-tailed macaque. *Primates*, 34, 301–309.

Mitchell, R.W., & Hamm, M. (1997). The interpretation of animal psychology: Anthropomophism or behavior reading? *Behaviour*, *134*, 173–204.

Mivart, St. G. (1873). On *Lepilemur* and *Cheirogaleus* and on the zoological rank of the *Lemuroidea*. *Proceedings of the Zoological Society of London 1873*, 484–510.

Morbeck, M.E., Galloway, A., & Zihlman, A.L. (Eds.) (1997). *The evolving female*. Princeton, NJ: Princeton University Press.

Mori, A. (1983). Comparison of the communicative vocalizations and behaviors of group ranging in Eastern gorillas, chimpanzees and pygmy chimpanzees. *Primates*, *24*, 486–500.

Mori, A. (1984). An ethological study of pygmy chimpanzees in Wamba, Zaire: A comparison with chimpanzees. *Primates*, *25*, 255–278.

Mori, U. (1979a). Individual relationships within a unit. In M. Kawai (Ed.), *Contributions to primatology, Vol 16*: *Ecological and sociological studies of gelada baboons*, (pp.93–123). Basel: S. Karger.

Mori, U. (1979b). Inter-unit relationships. In M. Kawai, (Ed.), *Contributions to primatology, Vol 16: Ecological and sociological studies of gelada baboons*, (pp.83–92). Basel: S. Karger.

Munn, C.A. (1986). Birds that cry "wolf". *Nature*, *319*, 143–145.

Muroyama, Y. (1991). Mutual reciprocity of grooming in female Japanese macaques (*Macaca fuscata*) *Behaviour*, *119*, 161–170.

Muroyama, Y. (1994). Exchange of grooming for allomothering in female patas monkeys. *Behaviour*, *128*, 103–119.

Muroyama, Y. (1996). Decision-making in grooming by Japanese macaques. *International Journal of Primatology*, *17*, 817–830.

Myers, D.G. (1983). *Social psychology*. New York: McGraw-Hill.

Nakamichi, M., Itorgawa, N., Imakawa, S., & Machida, S. (1995). Dominance relations among adult females in a free-ranging group of Japanese monkeys at Katsuyama. *American Journal of Primatology*, *37*, 241–251.

Napier, J.R., & Napier, P.H. (1967). *Handbook of living primates. Morphology, ecology and behaviour of nonhuman primates*. London: Academic Press.

Neel, M.A. (1996). Mirrors, monkeys and group dynamics. *Primates*, *37* (4), 411–421.
Nicolson, N. (1987). Infants, mothers and other females. In B.B. Smuts, D.L. Cheney, R.M. Seyfarth, R.W. Wrangham, & T.T. Struhsaker (Eds.), *Primate societies*, (pp.330–342). Chicago: University of Chicago Press.
Nishida, T. (1979). The social structure of chimpanzees of the Mahale Mountains. In D.A. Hamburg & E.R. McCown (Eds.) *The great apes*, (pp.73–121). Menlo Park, CA: Benjamin/Cummings.
Nishida, T. (1983). Alpha status and agonistic alliance in wild chimpanzees (*Pan troglodytes schweinfurthii*). *Primates*, *24*, 318–336.
Nishida, T. (1994). Review of recent findings on Mahale chimpanzees. In R.W. Wrangham, W.C. McGrew, F.B.M. de Waal, & P.G. Heltne (Eds.), *Chimpanzee culture*, (pp.373–396). Cambridge, MA: Harvard University Press.
Nöe, R. (1989). *Coalition formation among male baboons*. Netherlands: University of Utrecht, Ph.D. Dissertation.
Nöe, R. (1990). A veto game played by baboons: A challenge to the use of the prisoner's dilemma as a paradigm for reciprocity and cooperation. *Animal Behaviour*, *39*, 78–90.
Nöe, R. (1992). Alliance formation among male baboons: Shopping for profitable partners. In A.H. Harcourt, & F.B.M. de Waal (Eds.), *Coalitions and alliances in humans and other animals*. Oxford: Oxford University Press.
Nöe, R., van Schaik, C.P., & van Hooff, J.A.R.A.M. (1991). The market effect: An explanation of pay-off asymmetries among collaborating animals. *Ethology*, *87*, 87–97.
Ortony, A., & Turner, T.J. (1990). What is so basic about basic emotions? *Psychological Review*, *97*, 315–331.
Packer, C. (1977). Reciprocal altruism in olive baboons. *Nature*, *265*, 441–443.
Packer, C. (1980). Male care and exploitation of infants in *Papio anubis*. *Animal Behaviour*, *27*, 37–45.
Parker, S.T., & Milbrath, C. (1994). Contributions of imitation and role-playing games to the construction of self in primates. In S.T. Parker, R.W. Mitchell, & M. Boccia (Eds.), *Self-awareness in animals and humans. Developmental perspectives*, (pp.108–128). Cambridge: Cambridge University Press.
Parker, S.T., Mitchell, R.W., & Boccia M. (Eds.) (1994). *Self-awareness in animals and humans. Developmental perspectives*. Cambridge: Cambridge University Press.
Parker, S.T., & Russon, A.E. (1996). On the wild side of culture and cognition in the great apes. In A.E. Russon, K.A. Bard, & S.T. Parker (Eds.), *Reaching into thought. The minds of the great apes*, (pp.430–450). Cambridge: Cambridge University Press.
Passingham, R.E. (1982). *The human primate*. New York: W.H. Freeman & Company.
Patterson, F.G.P., & Cohn, R.H. (1994). Self-recognition and self-awareness in lowland gorillas. In S.T. Parker, R.W. Mitchell, & M.L. Boccia (Eds.), *Self-awareness in animals and humans: Developmental perspectives*, (pp.273–281). Cambridge: Cambridge University Press.
Pereira, M. (1995). Development and social dominance among group-living primates. *American Journal of Primatology*, *39*, 143–175.
Perloe, S.I. (1992). Male mating competition, female choice and dominance in a free ranging group of Japanese macaques. *Primates*, *33*, 289–304.
Perner, J. (1991a). On representing that: The asymmetry between belief and desire in children's theory of mind. In D. Frye & C. Moore (Eds.) *Children's theories of mind: Mental states and social understanding*, (pp.139–155). Hillsdale, NJ: Lawrence Erlbaum Associates.
Perner, J. (1991b). *Understanding the representational mind*. Cambridge, MA: MIT Press.
Perry, S. (1996). Female–female social relationships in wild white-faced capuchin monkeys (*Cebus capucinus*). *American Journal of Primatology*, *40*, 167–182.
Petersen, M.R. (1982). The perception of species-specific vocalization by primates: A conceptual framework. In C.T. Snowdon, C.H. Brown, & M.R. Petersen (Eds.), *Primate communication*, (pp.212–238). Cambridge: Cambridge University Press.

Phillips, K.A., Bernstein, I.S., Dettmer, E.L., Devermann, H., & Powers, M. (1994). Sexual behaviour in brown capuchins (*Cebus apella*). *Journal of Primatology*, *15*, 907–917.

Piaget, J. (1963). *The origins of intelligence in children.* New York: Norton.

Piaget, J., & Inhelder, B. (1956). *The child's conception of space.* London: Routledge & Kegan Paul.

Platt, M.M., & Thompson, R.L. (1985). Mirror responses in a Japanese macaque troop (Arashiyama West). *Primates*, *26*, 300–314.

Povinelli, D.J. (1994). A theory of mind is in the head, not in the heart. *Behavioural and Brain Sciences*, *17*, 571–584.

Povinelli, D.J., & Davis, D.R. (1994). Differences between chimpanzees (*Pan troglodytes*) and humans (*Homo sapiens*) in the resting state of the index finger: Implications for pointing. *Journal of Comparative Psychology*, *108*, 134–139.

Povinelli, D.J., Nelson, K.E., & Boysen, S.T. (1990). Inferences about guessing and knowing by chimpanzees (*Pan troglodytes*). *Journal of Comparative Psychology*, *104*, 203–210.

Povinelli, D.J., & Preuss, T.M. (1995). Theory of mind: Evolutionary history of a cognitive specialization. *Trends in Neurosciences*, *18*, 418–424.

Povinelli, D.J., Rulf, A.B., Landau, K.R., & Bierschwale, D.T. (1993). Self-recognition in chimpanzees (*Pan troglodytes*). Distribution, ontogeny and patterns of emergence. *Journal of Comparative Psychology*, *107*, 347–372.

Premack, D. (1976). *Intelligence in ape and man.* Hillsdale, NJ: Lawrence Erlbaum.

Premack, D. (1988). "Does the chimpanzee have a theory of mind?" revisited. In R.W. Byrne & A. Whiten (Eds.), *Machiavellian intelligence*, (pp.94–110). Oxford: Clarendon Press.

Premack, D., & Dasser, V. (1991). Perceptual origins and conceptual evidence for theory of mind in apes and children. In A. Whiten (Ed.), *Natural theories of mind. Evolution, development and simulation of everyday mindreading*, (pp.253–266). Oxford: Basil Blackwell.

Premack, D., & Premack, A. (1983). *The mind of an ape.* New York: Norton.

Premack, D., & Premack, A. (1994). Levels of causal understanding in chimpanzees and children. *Cognition*, *50*, 347–362.

Premack, D., & Woodruff, G. (1978). Does the chimpanzee have a theory of mind? *Behavioural and Brain Sciences*, *1*, 515–526.

Preuschoft, S. (1992). "Laughter" and "smile" in Barbary macaques (*Macaca sylvanus*). *Ethology*, *91*, 220–236.

Preuschoft, S. (1995). *"Laughter" and "smile" in macaques: An evolutionary perspective.* Netherlands: University of Utrecht, Ph.D. thesis.

Preuschoft, S., & Preuschoft, H. (1994). Primate non vocal communication: Our communicatory heritage. In W. Noth (Ed.), *Origins of semiosis*, (pp.61–100). Berlin: Mouton de Gruyter.

Preuschoft, S., & van Hooff, J.A.R.A.M. (1995). Homologizing primate facial displays: A critical review of methods. *Folia Primatologica*, *65*, 121–137.

Provine, R.R. (1986). Yawning as a stereotyped action pattern and releasing stimulus. *Ethology*, *72*, 109–122.

Provine, R.R. (1996). Contagious yawning and laughter. Significance for sensory feature detection, motor pattern generation, imitation and the evolution of social behaviour. In C.M. Heyes, & B.G. Galef Jnr. (Eds.), *Social learning of animals: The roots of culture*, (pp.179–208). San Diego, CA: Academic Press.

Provine, R.R. (1997). Yawns, laughs, smiles, tickles and talking: Naturalistic and laboratory studies of facial action and social communication. In J.A. Russell, & J.M. Fernandez Dols (Eds.), *The psychology of facial expression*, (pp.158–175). Cambridge: Cambridge University Press.

Provine, R.R., & Fischer, K.R. (1989). Laughing, smiling and talking: Relation to sleeping and social context in humans. *Ethology*, *83*, 295–305.

Provine, R.R., & Young, Y.L. (1991). Laughter: A stereotyped human vocalization. *Ethology*, *89*, 115–124.

Ransom, T.W. (1981). *Beach troop of the Gombe*. Lewisburg: Bucknell University Press.

Redican, W.K. (1975). Facial expressions in nonhuman primates. In L.A. Rosenblum (Ed.), *Primate behavior, 14*, (pp.104–194). New York: Academic Press.

Reynolds, P.C. (1981). *On the evolution of human behavior: The argument from animals to man*. Berkeley, CA: University of California Press.

Reynolds, V. (1961). *Social life of a colony of rhesus monkeys* (*Macaca mulatta*). London: London University, Ph.D. Thesis.

Reynolds, V. (1976). The origins of a behavioural vocabulary: The case of the rhesus monkey. *Journal for the Theory of Social Behaviour, 6*, 105–142.

Reynolds, V. (1984). *The biology of human action* (2nd ed.). San Francisco: W.H. Freeman.

Reynolds, V., & Reynolds, F. (1965). Chimpanzees in the Budongo Forest. In I. DeVore (Ed.), *Primate behavior: Field studies of monkeys and apes*, (pp.368–424). New York: Holt, Rinehart & Winston.

Rhine, R.J. (1975). The order of movement of yellow baboons (*Papio cynocephalus*). *Folia Primatologica, 23*, 72–104.

Rhine, R.J., & Westlund, B.J. (1981). Adult male positioning in baboon progressions: Order and chaos revisited. *Folia Primatologica, 35*, 77–116.

Richman, B. (1976). The synchronization of voices by gelada monkeys. *Primates, 19*, 569–581.

Richman, B. (1987). Rhythm and melody in gelada vocal exchanges, *Primates, 28*, 199–223.

Rinn, W.E. (1984). The neuropsychology of facial expressions: A review of the neurological and psychological mechanisms for producing facial expressions. *Psychological Bulletin, 95*, 52–77.

Rodman, P.S., & Mitani, J.C. (1987). Orangutans: Sexual dimorphism in a solitary species. In B.B. Smuts, D.L. Cheney, R.M. Seyfarth, R.W. Wrangham, & T.T. Struhsaker (Eds.), *Primate societies*, (pp.146–154). Chicago: University of Chicago.

Rowell, T.E. (1966). Hierarchy in the organization of a captive baboon group. *Animal Behaviour, 14*, 430–443.

Rowell, T.E. (1972). *The social behaviour of monkeys*. Harmondsworth, UK: Penguin Books.

Rowell, T.E. (1973). Social organization of wild Talapoin monkeys. *American Journal of Physical Anthropology, 38*, 593–598.

Rowell, T.E. (1974). The concept of social dominance, *Behavioural Biology, 11*, 131–154.

Rowell, T.E. (1988). Beyond the one-male group. *Behaviour, 104*, 189–201.

Rowell, T.E., & Hinde, R.A. (1962). Vocal communication by the rhesus monkey (*Macaca mulatta*). *Proceedings of the Zoological Society of London, 138*, 279–294.

Rowell, T.E., Hinde, R.A., & Y. Spencer-Booth (1964). "Aunt"–infant interaction in captive rhesus monkeys. *Animal Behaviour, 12*, 219–226.

Rumbaugh, D.M. (1990). Comparative psychology and the great apes: Their competency in learning, language and numbers. *Psychological Record, 40*, 15–39.

Russell, J.A. (1994). Is there a universal recognition of emotion from facial expressions? *Psychological Bulletin, 115*, 102–141.

Russon, A.E. (1997). Exploiting the expertise of others. In A. Whiten & R.W. Byrne, (Eds.), *Machiavellian intelligence II*, (pp.174–206). Cambridge: Cambridge University Press.

Russon, A.E., & Bard, K.A. (1996). Exploring the minds of the great apes: Issues and controversies. In A.E. Russon, K.A. Bard, & S.T. Parker (Eds.), *Reaching into thought: The minds of the great apes*. Cambridge: Cambridge University Press.

Russon, A.E., & Galdikas, B.M.F. (1995). Constraints on great apes' imitation, model and action selectivity in rehabilitant orangutan (*Pongo pygmaeus*) imitation. *Journal of Comparative Psychology, 109*, 5–17.

Sade, D.S. (1972). Sociometrics of *Macaca mulatta*, 1: Linkages and cliques in grooming matrices. *Folia Primatologica, 18*, 196–223.

Sahakian, W.S. (1982). *History and systems of social psychology*. Washington: Hemisphere Publication Company.

Sapolsky, R.M. (1996). Why should an aged male baboon ever transfer troops? *American Journal of Primatology, 39*, 149–157.
Sarich, V.M. (1984). Pygmy chimpanzee systematics: A molecular perspective. In R.L. Susman (Ed.), *The pygmy chimpanzee: Evolutionary biology and behavior*, (pp.43–48). New York: Plenum Press.
Sarich, V.M., & Cronin, J.E. (1976). Molecular systematics of the primates. In M. Goodman, & R.E. Tashian (Eds.), *Molecular anthropology*, (pp.141–170). New York: Plenum Press.
Saunders, C.D., & Hausfater, G. (1988). The functional significance of baboon grooming behavior. *Annals of the New York Academy of Sciences, 525*, 430–432.
Savage-Rumbaugh, E.S. (1984). *Pan paniscus* and *Pan troglodytes*: Contrasts in pre-verbal communicative competence. In R.L. Susman (Ed.), *The pygmy chimpanzee: Evolutionary biology and behaviour*, (pp.395–413). New York: Plenum Press.
Savage-Rumbaugh, E.S. (1988). The functional significance of baboon grooming behavior. *Annals of the New York Academy of Sciences, 525*, 430–432.
Savage-Rumbaugh, E.S., & Wilkerson, B. (1978). Socio-sexual behaviour in *Pan paniscus* and *Pan troglodytes*: A comparative study. *Journal of Human Evolution, 7*, 327–344.
Savage-Rumbaugh, E.S., Wilkerson, B., & Bakeman, R. (1977). Spontaneous gestural communication among conspecifics in the pygmy chimpanzee (*Pan paniscus*). In G.H. Bourne (Ed.), *Progress in ape research*, (pp.96–116). New York: Academic Press.
Schjelderup-Ebbe, T. (1935). Social behaviour of birds. In C. Murchison (Ed.), *Handbook of social psychology*, (pp.947–972). Worcester, MA: Clark University Press.
Schneider, K., & Unzner, L. (1992). Preschoolers' attention and emotion in an achievement and affect game: A longitudinal study. *Cognition and Emotion, 6*, 37–63.
Schubert, J.N. (1991). Human vocalizations in agonistic political encounters. In G. Schubert, & R.D. Martin (Eds.), *Primate politics*, (pp.207–220). Carbondale: Southern Illinois University Press.
Schultz, A.H. (1938). The relative weight of the testes in primates. *Anatomical Record, 72*, 387–394.
Seyfarth, R.M. (1977). A model of social grooming among adult female monkeys. *Journal of Theoretical Biology, 65*, 671–698.
Seyfarth, R.M. (1981). Do monkeys rank each other? *Behavioral and Brain Sciences, 4*, 447–448.
Seyfarth, R.M. (1983). Grooming and social competition in primates. In R.A. Hinde (Ed.), *Primate social relationships*. Oxford: Blackwell Scientific.
Seyfarth, R.M. (1992). Meaning and mind in monkeys. *Scientific American, 267* (6), 122–128.
Seyfarth, R.M., & Cheney, D.L. (1984). Grooming, alliances and reciprocal altruism in vervet monkeys. *Nature, 308*, 541–543.
Seyfarth, R.M., & Cheney, D.L. (1988). Empirical tests of reciprocity theory: Problems in assessment. *Ethology and Sociobiology, 9*, 181–188.
Seyfarth, R.M., Cheney, D.L., Harcourt A.H., & Stewart K.J. (1994). The acoustic features of Gorilla double grunts and their relation to behavior. *American Journal of Primatology, 33*, 31–50.
Seyfarth, R.M., Cheney, D.L., & Hinde, R.A. (1978). Some principles relating social interaction and social structure among primates. In D.J. Chivers, & J. Herbert (Eds.), *Recent advances in primatology* (Vol. 1, pp.39–51). New York: Academic Press.
Short, R.V. (1979). Sexual selection and its component parts, somatic and genital selection, as illustrated by man and the great apes. In J.S. Rosenblatt, R.A. Hinde, C. Beer, & M.-C. Busnell (Eds.), *Advances in the study of behaviour, 9*, (pp.131–158). New York: Academic Press.
Silk, J.B. (1982). Altruism among female *Macaca radiata*: Explanations and analysis of patterns of grooming and coalition formation. *Behaviour, 79*, 162–188.
Simonds, P.E. (1973). Outcast males and social structure among bonnet macaques (*Macaca radiata*). *American Journal of Physical Anthropology, 58*, 599–604.
Simonds, P.E. (1974). *The social primates*. New York: Harper & Row.

Simpson, M.J.A. (1973). Social grooming of male chimpanzees. In J.H. Crook, & R.P. Michael (Eds.), *Comparative ecology and behaviour of primates*, (pp.411–502). New York: Academic Press.

Simpson, M.J.A. (1991). On declaring commitment to a partner. In P.P.G. Bateson (Ed.), *Development and integration of behaviour. Essays in honour of Robert Hinde*, (pp.271–293). Cambridge: Cambridge University Press.

Simpson, J.A., Gangestad, S.W., & Nations, C. (1996). Sociosexuality and relationship initiation: An ethological perspective of nonverbal behavior. In G.J.O. Fletcher, & J. Fitness (Eds.), *Knowledge structures in close relationships: A social psychological approach*, (pp.121–146). Mahwah, NJ: Lawrence Erlbaum Associates.

Slater, P.J.B. (1973). Describing sequences of behaviour. In P.P.G. Bateson, & H.P. Klopfer (Eds.), *Perspectives in ethology*. Vol. 1, (pp.131–153). New York: Plenum Press.

Small, M.F. (1990). Promiscuity in Barbary macaques (*Macaca sylvanus*). *American Journal of Primatology*, *20*, 267–282.

Small, M.F. (1993). *Female choices. Sexual behavior of female primates*. Ithaca: Cornell University Press.

Smith, D.G. (1993). A 15-year study of the association between dominance rank and reproductive success of male rhesus macaques. *Primates*, *34*, 471–480.

Smith, P.K. (1984). *Play in animals and humans*. Oxford: Blackwell.

Smith, W.J. (1965). Message, meaning and context in ethology. *American Naturalist*, *99*, 405–409.

Smith, W.J. (1977). *The behaviour of communicating*. Cambridge, MA: Harvard University Press.

Smith, W.J., Shields, W.E., Schull, J., & Washburn, D.A. (1997). The uncertain response in humans and animals. *Cognition*, *62*, 75–97.

Smuts, B.B. (1985). *Sex and friendship in baboons*. Chicago: Aldine.

Smuts, B.B. (1987). Gender, aggression and influence. In B.B. Smuts, D.L. Cheney, R.M. Seyfarth, R.W. Wrangham, & T.T. Struhsaker (Eds.), *Primate societies*. Chicago: University of Chicago Press.

Smuts, B.B. (1997). Social relationships and life histories of primates. In M.E. Morbeck, A. Galloway, & A.L. Zihlman (Eds.), *The evolving female*, (pp.60–75). Princeton, NJ: Princeton University Press.

Smuts, B.B., Cheney, D.L., Seyfarth, R.M., Wrangham, R.W., & Struhsaker, T.T. (Eds.) (1987). *Primate societies*. Chicago: University of Chicago Press.

Smuts, B.B., & Smuts, R.W. (1993). Male aggression and sexual coercion of females in non-human primates and other mammals: Evidence and theoretical implications. In P.J.B. Slater, J.S. Rosenblatt, C.T. Snowdon, & M. Milinski (Eds.), *Advances in the study of behaviour*, *22*, (pp.1–63). New York: Academic Press.

Smuts, B.B., & Watanabe, J.M. (1990). Social relationships and ritualized greetings in adult male baboons (*Papio cynocephalus anubis*). *International Journal of Primatology*, *11*, 147–172.

Snowdon, C.T. (1982). Linguistic and psycholinguistic approaches to primate communication. In C.T. Snowdon, C.H. Brown, & M.R. Petersen (Eds.), *Primate communication*, (pp.212–238). Cambridge: Cambridge University Press.

Snowdon, C.T. (1990). Mechanisms maintaining monogamy in monkeys. In D.A. Dewsbury (Ed.), *Contemporary issues in comparative psychology*, (pp.225–251). Sunderland, MA: Sinauer Associates.

Snowdon, C.T. (1993). The rest of the story: Grooming, group size and neotropical primates. *Behavioral and Brain Sciences*, *16*, 718.

Snowdon, C.T., & Cleveland, J. (1984). "Conversations" among pygmy marmosets. *American Journal of Primatology*, *7*, 15–20.

Sperling, S. (1991). Baboons with briefcases: Feminism, functionalism and sociobiology in the evolution of primate gender. *Signs: Journal of Women in Culture and Society*, *17*, 1–27.

Stammbach, E. (1988). Group responses to specially skilled individuals in a *Macaca fascicularis* group. *Behaviour, 107*, 241–266.

Stewart, K.J. (1988). Chronicles of the Pumphouse Gang. Review of *Almost human. New York Times Book Review.* (10 January), 14.

Stewart, K.J., & Harcourt, A.H. (1994). Gorillas' vocalizations during rest periods: Signals of impending departure. *Behaviour, 130*, 29–40.

Struhsaker, T.T. (1967). Auditory communication among vervet monkeys (*Cercopithecus aelthiops*). In S.A. Altmann (Ed.), *Social communication among primates*, (pp.281–324). Chicago: University of Chicago Press.

Struhsaker, T.T. (1975). *The red colobus monkey.* Chicago: University of Chicago Press.

Strum, S.C. (1987). *Almost human. A journey into the world of baboons.* New York: Random House.

Strum, S.C., Forster, D., & Hutchins. E. (1997). Why Machiavellian intelligence may not be Machiavellian. In A. Whiten, & R.W. Byrne (Eds.), *Machiavellian intelligence II*, (pp.50–85). Cambridge: Cambridge University Press.

Sugiyama, Y. (1969). Social behavior of chimpanzees in the Budongo Forest, Uganda. *Primates, 10*, 197–225.

Susman, R.L. (1984). *The pygmy chimpanzee. Evolutionary biology and behavior.* New York: Plenum Press.

Symmes, D., & Biben, M. (1992). Vocal development in nonhuman primates. In H. Papousek, U. Jurgens, & M. Papousek (Eds.), *Nonverbal vocal communication. Comparative and developmental approaches*, (pp.123–140). Cambridge: Cambridge University Press.

Tanaka, M. (1996). Information integration about object–object relationships by chimpanzees (*Pan troglodytes*). *Journal of Comparative Psychology, 110*, 323–335.

Taub, D. (1980). Female choice and mating strategies among wild Barbary macaques (*Macaca sylvanus L*). In D.G. Lindburg (Ed.), *The macaques: Studies in ecology, behaviour and evolution*, (pp.287–344). New York: Van Nostrand Rheinhold.

Thierry, B., Demaria, C., Preuschoft S., & Desportes, C. (1989). Structural convergence between silent bared-teeth display and relaxed open-mouth display in the Tonkean macaques (*Macaca tonkeana*). *Folia Primatologica, 52*, 178–184.

Thomas, J. (1995). *Meaning in interaction: An introduction to pragmatics.* London: Longman.

Thompson, R.K.R. (1995). Natural and relational concepts in animals. In H.L. Roitblat & J.-A. Meyer (Eds.), *Comparative appproaches to cognitive science*, (pp.175–224). Cambridge, MA: MIT Press.

Tinbergen, N. (1963). On the aims and methods of ethology. *Zeitschrift für Tierpsychologie, 20*, 110–433.

Todt, D. (1987). *Duels, duets, dialogues: On communication processes that mediate social interactions.* Plenary paper presented at the 20th International Ethology Conference, Madison, Wisconsin.

Tomasello, M. (1996). Do apes ape? In C.M. Heyes, & B.G. Galef Jnr. (Eds.), *Social learning of animals: The roots of culture.* San Diego: Academic Press.

Tomasello, M., & Call, J. (1994). The social cognition of monkeys and apes. *Yearbook of Physical Anthropology, 37*, 273–305.

Tomasello, M., & Call, J. (1997). *Primate cognition.* New York: Oxford University Press.

Tomasello, M., Call, J., Nagell, K., Olguin, R., & Carpenter, M. (1994). The learning and use of gestural signals by young chimpanzees: A trans-generational study. *Primates, 35*, 137–154.

Trivers, R. (1971). The evolution of reciprocal altruism. *Quarterly Review of Biology, 46*, 35–57.

Tutin, C.E.G. (1979). Mating patterns and reproductive strategies in a community of wild chimpanzees (*Pan troglodytes schweinfurthii*). *Behavioral Ecology and Sociobiology, 6*, 39–48.

van Hooff, J.A.R.A.M. (1962). Facial expressions in higher primates. *Symposium of the Zoological Society of London, 8*, 97–125.

van Hooff, J.A.R.A.M. (1967). The facial displays of the catarrhine monkeys and apes. In D. Morris (Ed.), *Primate ethology*, (pp.7–68). Chicago: Aldine.

van Hooff, J.A.R.A.M. (1972). A comparative approach to the phylogeny of laughter and smiling. In R.A. Hinde (Ed.), *Non-verbal communication*, (pp.209–241). Cambridge: Cambridge University Press.

van Hooff, J.A.R.A.M. (1973). A structural analysis of the social behaviour of a semi-captive group of chimpanzees. In M. von Cranach, & I. Vine (Eds.), *Social communication and movement*, (pp.75–162). London: Academic Press.

Vasey, P.L. (1995). Homosexual behaviour in primates: A review of evidence and theory. *International Journal of Primatology*, *16*, 173–204.

Visalberghi, E., & Fragaszy, D. (1990). Do monkeys ape? In S.T. Parker & K.T. Gibson (Eds.), *Language and intelligence in monkeys and apes*. New York: Cambridge University Press.

Visalberghi, E., & Limongelli, L. (1994). Lack of comprehension of cause–effect relations in tool-using Capuchin monkeys (*Cebus apella*). *Journal of Comparative Psychology*, *108*, 15–22.

Wallman, J. (1992). *Aping language*. Cambridge: Cambridge University Press.

Washburn, S.L., & DeVore, I. (1963). The social life of baboons. In C.H. Southwick (Ed.), *Primate social behaviour*, (pp.98–113). New York: Van Nostrand Reinhold.

Watson, J.B. (1908). Imitation in monkeys. *Psychological Bulletin*, *1*, 169–178.

Watts, D.P. (1995a). Post-conflict social events in wild mountain gorillas (*Mammalia, Hominoidea*) I. Social interactions between opponents. *Ethology*, *100*, 139–157.

Watts, D.P. (1995b). Post-conflict social events in wild mountain gorillas II. Redirection, side direction and consolation. *Ethology*, *100*, 158–174.

White, F.J. (1992a). Activity budgets, feeding behaviour and habitat use of pygmy chimpanzees at Lomoko, Zaire. *American Journal of Primatology*, *26*, 215–223.

White, F.J. (1992b). Pygmy chimpanzees' social organization: Variations with party size and between study sites. *American Journal of Primatology*, *26*, 203–214.

White, F.J. (1996). *Pan paniscus* 1973 to 1996: Twenty-three years of field research. *Evolutionary Anthropology*, *5*, 11–17.

Whiten, A. (1991). The emergence of mindreading. Steps towards an interdisciplinary enterprise. In A. Whiten (Ed.), *Natural theories of mind. Evolution, development and simulation of everyday mindreading*, (pp.319–331). Oxford: Basil Blackwell.

Whiten, A. (1993). Evolving a theory of mind: The nature of non-verbal mentalism in other primates. In S. Baron-Cohen, H. Tager-Flusberg, & D.J. Cohen (Eds.), *Understanding other minds. Perspectives from autism*, (pp.367–396). Oxford: Oxford University Press.

Whiten, A. (1994). Grades of mindreading. In C. Lewis, & P. Mitchell (Eds.), *Children's early understanding of mind. Origins and development*, (pp.47–70). Hove, UK: Lawrence Erlbaum Associates Ltd.

Whiten, A. (1996a). Ape mind, monkey mind. *Evolutionary Anthropology*, *5*, 3–4.

Whiten, A. (1996b). Imitation, pretense and mind reading: Secondary representation in comparative psychology and developmental psychology? In A.E. Russon, K.A. Bard, & S.T. Parker (Eds.), *Reaching into thought. The minds of the great apes*, (pp.300–324). Cambridge: Cambridge University Press.

Whiten, A., & Byrne, R.W. (1988a). Tactical deception in primates. *Behavioral and Brain Sciences*, *11*, 233–272.

Whiten, A., & Byrne, R.W. (1988b). The St. Andrews catalogue of tactical deception in primates. St. *Andrews Psychological Report*, No. 10. St. Andrews, UK: St. Andrews University.

Whiten, A., & Byrne, R.W. (1997). *Machiavellian Intelligence II*. Cambridge: Cambridge University Press.

Whiten, A., & Perner, J. (1991). Fundamental issues in the multidisciplinary study of mindreading. In A. Whiten (Ed.), *Natural theories of mind: Evolution, development and simulation of everyday mindreading* (pp.1–17). Oxford, UK: Basil Blackwell.

Wrangham, R.W. (1980). An ecological model of female bonded primate groups. *Behaviour, 75,* 262–300.

Wrangham, R.W. (1987). Evolution of social structure. In B.B. Smuts, D.L. Cheney, R.M. Seyfarth, R.W. Wrangham, & T.T. Struhsaker (Eds.), *Primate societies* (pp.282–296). Chicago: University of Chicago Press.

Wrangham, R.W. (1993). The evolution of sexuality in chimpanzees and bonobos. *Human Nature, 4,* 47–79.

Yamagiwa, J. (1987). Intra- and inter-group interactions of an all-male group of Virunga mountain gorillas (*Gorilla gorilla beringei*). *Primates, 28,* 1–30.

Yamagiwa, J. (1992). Functional analysis of social staring behaviour in an all-male group of mountain gorillas. *Primates, 33,* 523–524.

Zeifman, D., & Hazan, C. (1997). Attachment: The bonds in pair-bonds. In J.A. Simpson, & D.T. Kenrick (Eds.), *Evolutionary social psychology,* (pp.237–263). Mahwah, NJ: Lawrence Erlbaum Associates.

Zeller, A.C. (1980). Primate facial gestures. *International Journal of Human Communication, 13,* 565–606.

Zeller, A.C. (1987). Communication by sight and smell. In B.B. Smuts, D.L. Cheney, R.M. Seyfarth, R.W. Wrangham, & T.T. Struhsaker (Eds.), *Primate societies,* (pp.433–439). Chicago: University of Chicago Press.

Zeller, A.C. (1991). Grooming interactions over infants in four species of primate. *Visual Anthropology, 5,* 63–86.

Zihlman, A.L. (1997). Natural history of apes: Life history features in females and males. In M.E. Morbeck, A. Galloway, & A.L. Zihlman (Eds.), *The evolving female,* (pp.86–103). Princeton, NJ: Princeton University Press.

Zimmerman, E., Newman, J.D., & Jurgens, U. (Eds.) (1995). *Current topics in primate vocal communication.* New York: Plenum.

Author index

Subject index